THE STANDARD FORM OF BUILDING CONTRACT

JCT 80

Second edition

JOHN PARRIS
LLB(Hons), PhD

COLLINS
8 Grafton Street, London W1

Collins Professional and Technical Books
William Collins Sons & Co. Ltd
8 Grafton Street, London W1X 3LA

First Edition published in Great Britain by
Granada Technical Books 1982
Reprinted 1983, 1984
Second Edition published by
Collins Professional and Technical Books 1985

Distributed in the United States of America
by Sheridan House, Inc.

British Library Cataloguing in Publication Data
Parris, John
The standard form of building contract: JCT 80.
— 2nd ed.
1. Joint Contracts Tribunal. Standard form of
building contract. 1980. 2. Building—Contracts
and specifications—Great Britain
I. Title
692'.8 TH425

ISBN 0-00-383027-6

Printed and bound in Great Britain by Mackays of Chatham, Kent

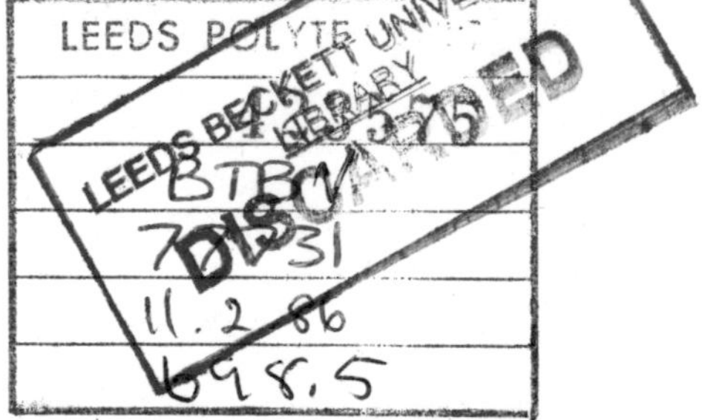

Acknowledgments
Extracts from JCT 80, Forms TNS/2, NSC/2, NSC/4, JCT Design and Build Form, JCT 63 and the JCT Guide are reproduced with the kind permisson of RIBA Publications Ltd, the copyright holder.

Clauses from the 'Blue' Form for Non-Nominated Sub-Contractors and the 'Green' Form for Nominated Sub-Contractors are reproduced with the permission of the Building Employers Confederation.

Contents

Preface to the second edition

The Standard Form of Building Contract JCT 80 is still far short of universal acceptance by employers. Contractors on the whole are quite happy with it, as are sub-contractors with the accompanying sub-contract documents. Some local authorities are very happy with it.

Clause 19 [see section 5.15] which allows the main contractor to select one of three sub-contractors nominated by the employer, has proved particularly useful. Contrary to many expectations, contractors have not selected the most expensive sub-contractor tenderer, but the one they consider will give the best performance and the least problems. Contractors know a great deal more about sub-contractors than any employer ever can.

In spite of this, some London boroughs and public authorities in Northern Ireland have abandoned JCT 80 for amended versions of GC/Works/1. This is a contract heavily biased in the employer's favour which, when Government departments use it, works well for three reasons: firstly, because the Government departments adopt a fair and even generous attitude to contractors and do not rely on the strict terms of the contract; secondly, because the select band of contractors who are invited to tender under it are not only efficient, but also a happy band of brothers who effectively decide among themselves who shall get what contract and at what price; thirdly, because these contractors are concerned not only with the current contract but all the future ones and it is in their interest to keep the employer sweet. Hence, the almost complete absence of litigation on GC/Works/1. None of these factors is likely to apply when a local or public authority, other than a Government department, uses the form and they are likely therefore to pay heavily for their choice in tender price and subsequent litigation.

In Scotland, employers are rejecting JCT 80 in favour of JCT 63. (The Scottish Building Committee have also, incidentally, refused to prepare a Scottish version of the JCT Intermediate Form IFC 84.)

Several architects have praised section 1.07 of this book and have said how it has helped them to persuade employers not to use JCT 63. I have therefore in this edition augmented that section by providing other reasons why that form should not be used. I entirely endorse what is said by Vincent Powell-Smith and John Sims in their useful book *Contract Documentation for Contractors* (1984): 'Since the issue of JCT 80, any professional adviser who recommends his client to use JCT 63 terms is guilty of professional negligence.' If the employer insists on using it, the professional should, to protect himself, put in writing his emphatic disagreement and thus ensure that it is the employer alone who bears the additional cost and inconvenience, which can well involve paying massive fluctuation sums on a delayed completion and loss of the right to claim liquidated damages.

It is also ironic that in the very year that the British Property Federation became full members of the Joint Contracts Tribunal, it should sponsor a revised version of the inequitable so-called 'standard contract' prepared by a small body, the Association of Consultant Architects. This is a form, like the traditional wartime case of sardines, for buying and selling, but not for use.

Apart from all this, major contracts have been executed recently in the United Kingdom on the FIDIC Form for international civil engineering work (based on the obsolete 4th Edition of the ICE Conditions) and on various United States standard forms.

The proliferation of different contracts is greatly to be deplored, for a multitude of reasons. Fundamentally JCT 80 is a good contract, fair to both employer and contractor, and without the 'calculated ambiguity' of JCT 63 which has led to so much litigation. Unhappily, it is not written in the most lucid of English, and it fails to distinguish between procedure which ought to be good practice and that which ought to be a contractual obligation. But the whole industry, from employers to sub-contractors, ought to stand behind it. It is their contract.

I have never disguised my admiration for the system of construction prevailing in the United States and Canada where, without bills of quantities, buildings are erected in shorter time, at half the cost (in spite of the high price of labour and materials) and with far fewer problems. Much of this is due to the system of one hundred per cent bonding of contractors and sub-contractors. But it is also due to the fact that banks are prohibited from writing construction bonds and bonding companies have found it much cheaper, where the contractor is in default, to take over the performance of a contract with their appointed management agents.

This system seems impossible to import into the United Kingdom and we have to deal with people as they are, not as we would wish them to be.

A disillusioned Tory MP once referred to a Tory prime minister as 'the best we have'. So far as contracts, unlike prime ministers, are concerned, employers have a choice. Unhesitatingly, I would describe JCT 80 as 'the best we have', for both public and private work.

It has been said that 'the JCT contracts are like a camel — a horse designed by a committee'. A camel, it should not be forgotten, is ideally suited to its environment, having among other things in its nostrils air filtering and refrigeration.

For this second edition, I have been greatly assisted by comments, advice and criticism made by a number of people including Tony Brett-Jones, Hugh Clamp, Frank Eaglestone, His Honour Edgar Fay QC, Peter Hibberd, Neil Jones, Philip King, Malcolm London, Owen Luder, Vincent Powell-Smith, John Sims, John Townsend, and John Uff QC.

Not every suggestion or criticism I have been able to accept, and the text, of course, remains my sole responsibility.

John Parris
April 1985

Chapter 1

Background to the contract

1.01 Earlier standard forms

In the last quarter of the nineteenth century, a standard form of contract for use in building work in the United Kingdom came into being. A copy of it, printed in *Hudson's Building Cases* (3rd Edition) Vol. 2 at page 632, shows that it consisted of some thirty clauses only.

From 1903, a development of this standard form became known as 'the RIBA Contract', a title that it retained until 1977 when, following severe judicial criticism and some acrid correspondence in *The Times* newspaper, the Royal Institute of British Architects (RIBA) withdrew its name from the document (although it claims copyright in the 1980 version). On 5 August 1977, judges and other court officials were told by item B586 of *Court Business* to refer to the Standard Form in future as 'the JCT Contract' or 'the Standard Form of Building Contract' and not as 'the RIBA Contract'. It is an injunction which is still being ignored by the judiciary and lawyers, other than the Official Referees.

1.02 The Joint Contracts Tribunal

From 1903, the Standard Form of Building Contract was a compilation made by a tripartite body consisting of representatives of the RIBA, the Building Employers Confederation (BEC), as it is now called, and the Institute of Building (IOB), as it then was. In 1931, the IOB withdrew, so that body henceforth was a 'Joint Contracts' one, consisting of the RIBA and the National Federation of Building Trades Employers, now the BEC.

There was a substantial rewrite of the Standard Form in 1939 and in 1963.

In 1952, the Royal Institution of Chartered Surveyors (RICS) became involved and by the year 1963, the Joint Contracts Tribunal consisted of representatives of ten bodies in the construction industry. Bodies representing sub-contractors subsequently joined.

In 1980, a resolution before the RIBA Council that the Institute should withdraw entirely from the Joint Contracts Tribunal, which had drafted the new version of the Standard Form, was defeated by only one vote. Disquiet had been caused by a JCT draft of another form, the Standard Form with Contractor's Design, which made no reference whatsoever to the employment of a qualified architect. The version finally adopted by the JCT, with which this book is not concerned, creates no obligation on the contractor to employ a qualified architect to do the designing, supervision or certification of sums due to the contractor.

However, architects now seem reconciled to the 'Design and Build' form, as it is now known, and it appears to have led to the employment of more architects rather than fewer, although perhaps not on such favourable terms.

1.03 The present JCT

The present body responsible for drafting the current form, therefore, is broadly representative of public sector employers, architects, contractors and sub-contractors.

The constituent bodies are:

Royal Institute of British Architects
Building Employers Confederation
Royal Institution of Chartered Surveyors
Association of County Councils
Association of Metropolitan Authorities
Association of District Councils
The Greater London Council
Committee of Associations of Specialist Engineering Contractors
Federation of Associations of Specialists and Sub-contractors
Association of Consulting Engineers
Scottish Building Contracts Committee
British Property Federation

The Joint Contracts Tribunal therefore is no longer a joint body and has never been a tribunal. There is a standing drafting committee responsible for the actual wording of the contracts. The drafts of this committee are circulated to members, and through them to their constituent bodies, and are not adopted unless all members are unanimous.

1.04 The range of contracts

It is the intention of the JCT to provide standard forms for every range of construction activity from package deals, contractors' design and build, cost plus to management fee. This intention can be best expressed diagrammatically (Figure 1).

This book deals solely with what may be termed the 'standard' Standard Form of Building Contract, i.e. that with quantities, in the local authority or private version, which are those intended for use for major works. For brevity this will be referred to as JCT 80.

Traditionally, the Local Authorities Form is regarded as the basic one for comment, with the Private Edition having minor variants. The two have substantially the same wording, though the Private

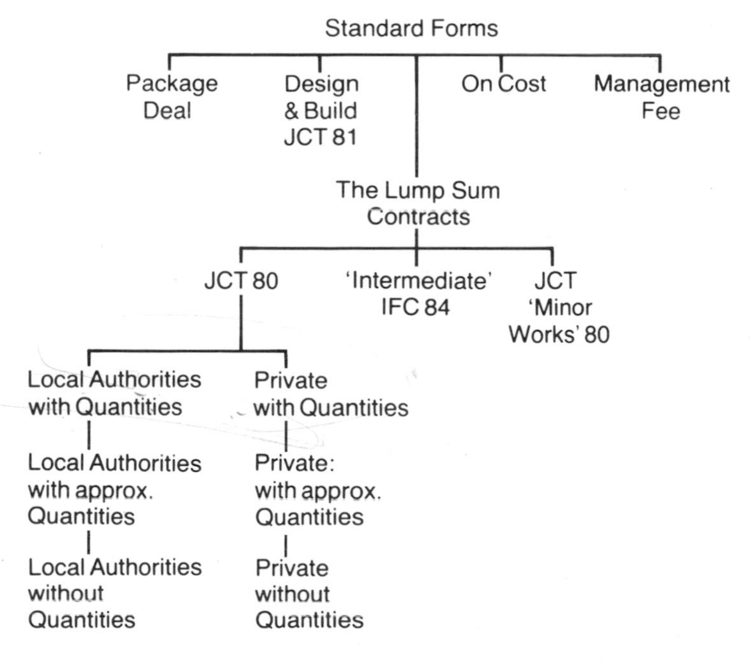

Figure 1

Edition omits Clause 19A dealing with fair wages, and sub-clause 30.5.3, dealing with the handling of retention monies [6.10], is omitted from the Local Authorities Edition. There are other differences.

Readers will find the other two 'lump sum' contracts dealt with in *A Commentary on the JCT Intermediate Form of Building Contract* by Neil Jones and David Bergman (1985), and Hugh Clamp's *The Shorter Building Contract Forms* (1984).

The 1963 edition of the Standard Form was a substantial rewrite, with minor amendments made almost every year until 1977.

Sub-contract documents formerly in use with JCT 63 were the 'Green Form' for nominated sub-contractors and the 'Blue Form' for non-nominated or, as they are now termed, 'domestic' sub-contractors. Both documents were issued jointly by the National Federation of Building Trades Employers (as it was then termed) and the Federation of Associations of Specialists and Sub-contractors and approved by other bodies.

The Joint Contracts Tribunal have replaced the 'Green Form' with NSC/4 or 4a and NFBTE/FASS replaced the 'Blue Form' with DOM/1, for use with the JCT 80 contract.

1.05 Bills of quantities

The most commonly used version form of the Standard Form is for use with bills of quantities. This is a peculiarly British practice and since it is largely unknown elsewhere or to most lawyers, it is necessary to describe it.

Bills were described in the Simon Report of 1944 *The Placing and Management of Building Contracts* as putting 'into words every obligation or service which will be required in carrying out the building project'.

British contractors are dedicated to it, so much so that the National Federation of Building Trades Employers (now the Building Employers Confederation) at one time had an agreement that no member would tender for work exceeding £8000 in value without bills of quantities being used. This agreement was held in *re Birmingham Association of Building Trade Employers' Agreement* (1963) to be contrary to the public interest and a violation of the Restrictive Trade Practices Act 1956.

Under this system, before going out to tender it is necessary for the architect to prepare his drawings in sufficient detail to enable a quantity surveyor to 'take off' (that is, measure) from them the actual amounts of work to be executed, often sub-divided into the various trades. This bill of quantities normally starts with what are termed the 'preliminaries', meaning items of expenditure which relate to the project as a whole, followed by the itemised bills.

When these are ready, contractors are invited to tender on the basis of these bills of quantities and to insert the price they require, not only for the whole project but also for each item of work.

In theory, this is a system which obviates the need for each contractor tendering to work out for himself, as contractors in the United States and Canada have to do, the quantities of material and work required. In practice, it does not work quite that way: hence the dedication of contractors to the system.

In the first place, the architect's drawings are rarely in sufficient detail to enable a bill of quantities to be prepared with total accuracy; so that any further detail has to be treated under the contract as an architect's instruction and a variation for which the contractor is entitled to receive extra remuneration [6.20].

Secondly, the art of evaluating from drawings the exact amount of materials and work required varies from the impossible, as in the case of excavations, to the difficult, as in the case of an air-conditioned computer room.

The bill of quantities in theory also purports to perform other functions: namely to determine the way in which the inevitable variations shall be priced [6.20]; to enable a fair valuation of the work done to be made for process payments made on interim certificates [6.02], and a complete revaluation of the final contract price for the final certificate. It is often believed that the final certificate is to be no more than the sum total of interim certificates issued plus any uncertified work. This is not the contract provision [11.07].

Practice does not coincide with theory. At one time early items shown in the bill of quantities were often 'front-loaded' by contractors so as to transfer as much of the employer's money to the contractor as soon as possible. This could result in later items being executed below cost and in desperate efforts to recoup from the employer enough by way of claims to make them profitable, or, in extreme cases, to take advantage of the highly favourable terms on which contractors can determine the contract [9.06].

With the advent of inflation and the general use of fluctuation clauses, contractors quickly realised that it could be more profitable to 'back-load' the tender. In that way, they could buy in materials early which they knew they would need at a later date and charge them out to the employer later at an enhanced price. In one case, a civil engineering contractor confessed that he had fabricated steel work four years before it was required and thereby quadrupled his profit on the work. Careful calculations had been done on interest charges and anticipated inflation of wages and materials. Every contractor employs a staff of quantity surveyors to sniff out the items where the quantity surveyor who prepared the bills has under-estimated the quantities, so that these can be highly priced in the tender by the contractor, and those where he has over-estimated the quantities, so that these can be under-priced.

There is a Standard Form of Building Contract without Quantities. The general view in the industry is that contractors will load their tenders when quantities are not used so that uneconomic prices will be quoted. In the current competitive situation that may not necessarily be so. The Minor Works Agreement is designed for use without bills of quantities and contractors show no great reluctance to tender for work under it. The JCT Intermediate Form, IFC 84, can also be used with or without bills of quantities. A thorough discussion of the alternatives will be found in Jones and Bergman: *A Commentary on the JCT Intermediate Form of Building Contract* (1985).

1.06 The Standard Method of Measurement

In order that there might be some resemblance of similitude in the way in which bills of quantities are prepared, the RICS and NFBTE, as it then was, prepared what is termed the 'Standard Method of Measurement', currently in its sixth edition—SMM6. There are, after all, no less than six ways in which a hole in the ground caused by excavation can be measured, as became apparent in the case of *Farr* v. *Ministry of Transport* (1965) on the Institution of Civil Engineers' contract, the ICE form.

JCT 63 (as revised) provided in clause 12(1) that:

> 'the quality and quantity of work included in the Contract Sum shall be deemed to be that set out in the Contract Bills, which Bills . . . shall be deemed to have been prepared in accordance with the principles of the Standard Method of Measurement 6th Edition . . .'

That was a masterly piece of ambiguity. Are 'the principles' of SMM6 the same as the detailed directions? Obviously not. Two 'deemed' provisions were enough to confuse any judge. 'Deemed' normally means that circumstances are to be treated as existing even if they manifestly are not.

JCT 80 clause 14.1 reads:

> 'the quality and quantity of the work included in the Contract Sum shall be deemed to be that which is set out in the Contract Bills.'

However, JCT 80 clause 2.2.2.1 reads:

> 'the Contract Bills . . . are to have been prepared in accordance with the Standard Method of Measurement of Building Works, 6th Edition . . .'

That is a bit more like English.
Clause 2.2.2.2 provides that:

> 'any departure from the method of preparation . . . or any error in description or in quantity or omission of items . . . shall not vitiate this Contract but . . . shall be corrected . . . [and treated as a variation].'

Since it is virtually impossible to accurately reduce the minutiae of building operations into words, there must inevitably be ambiguities in both SMM6 and in any bill of quantities.

Works not covered by SMM6 or the bills are frequent grounds of claims by the contractor, but the legal position is that the employer is not obliged to pay for 'things that everybody must have understood are to be done but which happen to be omitted from the quantities': *per* Channell J in *Patman and Fotheringham Ltd* v. *Pilditch* (1904).

In *Williams* v. *Fitzmaurice* (1858) there was a contract to build a house in accordance with a specification for a fixed price, but the specification omitted any reference to floorboards. It was held that the builder was not entitled to extra payment for the floorboards because it was evident from the specification that the builder was to do the flooring.

This principle does not appear to be known to architects and quantity surveyors these days so that, in reliance on the words of the contract, any sort of omission from SMM6 or the bills of quantities is treated as a variation, even though the operation not described is self-evident and necessary for the work which is described.

However, the provisions of SMM6 regarding site conditions seem to virtually guarantee the contractor successful claims regarding these [10.09].

A new Standard Method of Measurement, SMM7, was in preparation in 1985.

1.07 Why JCT 63 should not continue to be used

In 1981, RIBA Publications Ltd, without the authority of the JCT, and much to its annoyance, printed and published a further edition of the old 1963 edition. The President of the RIBA was forced to give an undertaking to the JCT that there was to be no further reprint without their consent. Despite this, quite a number of architects and their employers continue to use the 1963 version rather than the 1980 edition.

It is hoped that architects will not succumb to this temptation and that employers will not accept any suggestion from their professional advisers to use the old form. The reasons are quite apparent to those who devote any thought to the matter.

JCT 80 is a form agreed upon by the whole industry. When a contractor enters into the contract, he is not entering into a contract on 'the other' written standard terms of business', for the purpose of section 3 of the Unfair Contract Terms Act 1977. He is entering upon a contract which is as much his as his employer's.

If the obsolete 1963 form is used, it cannot be contended that it is anything other than the *employer's* 'written standard terms of business'. Moreover, since it is the employer's contract and not that of the industry, every provision in it must be construed *contra proferentem*, since the representatives of the construction industry have agreed to abandon it. This means that in the case of ambiguities in the 1963 edition (and there are many) a court or arbitrator is obliged to adopt the meaning most hostile to the employer. Any employer who uses the 1963 form makes it his own document as much as if he had had one specially drafted for him.

So far as standard contracts are concerned, Lord Pearson said in *Tersons* v. *Stevenage District Council* (1963) which dealt with the ICE Conditions (2nd edition):

> '[Counsel for the contractors] has contended that the maxim *verba fortius accipiuntur contra proferentem* should be applied in this case in favour of the contractor against the corporation, on the ground that the General Conditions were included in the invitation to tender sent by the corporation to the contractor.
>
> In my opinion, the maxim has little, if any, application in this case.
>
> The General Conditions are not a partisan document or an "imposed standard contract", as the phrase is, which is sometimes used.
>
> It is not drawn up by one party in its own interests and imposed upon the other party.
>
> It is a general form, evidently in common use, prepared and revised jointly by several representative bodies . . .'

The same was said, perhaps with less reason, of the Green Form in *County and District Properties* v. *Jenner* (1974).

All this is true of JCT 80. It is no longer true of JCT 63, and that must therefore inevitably be held to be the employer's form, with all the consequences. If there are any ambiguities or words that are difficult to interpret, they must always have such a construction placed upon them which is most hostile to the person that has produced the form. That is, it will now be construed *against* the employer.

The *contra proferentem* rule in relation to liquidation damages and extension of time clauses is discussed further on [7.01].

Fluctuations

Quite apart from that, there are substantial disadvantages to an employer in using the JCT 63 form.

Under JCT 80 a contractor who fails to complete the work on the due date (being either the original date for completion, or a date substituted for it by reason of extensions of time granted by the architect under clause 25) becomes liable to pay liquidated damages at the agreed rate to the employer.

But under the JCT 80 contract, whichever fluctuation clause is used [6.21], he also forfeits all right to receive increases in the cost of wages or materials which occur after the due date for completion.

If the clause 38 provisions for fluctuations are adopted, 38.4.7 provides that:

> 'no amount shall be added to . . . the computation of the amount stated as due in an Interim Certificate or in the Final Certificate . . . if the event . . . occurs after the Completion Date.'

If clause 39 is adopted, 39.5.7 makes similar provision. And if the price adjustment formula is used, clause 40 provides a similar effect, in rather different words, in 40.7.1.1.

In other words, in all cases the fluctuations are frozen at the due date for completion. If the work is delayed, the contractor has to bear any additional costs including, for example, wage rises promulgated after the date for completion.

The position under JCT 63 is quite different, apart from the situation where the 'Formula' provision is used.

Even if the contractor is years late in completion, he is entitled to fluctuations to the bitter end. As Lord Justice Salmon said in *Peak* v. *McKinney* (1970), dealing with a home-made Liverpool contract which was to the same effect as JCT 63: 'The corporation was liable to pay the plaintiffs the increase in wage rates right up to the moment the block was completed.' And Lord Justice Edmund Davies in the same case was equally categorical: 'The operation of the fluctuation clause is not to be regarded as impliedly restricted [to the completion date] but it extended throughout the whole period until the construction work was completed.'

One can therefore envisage a situation in which the contractor working under JCT 63 is two years late in completion (by no means unknown) and succeeds in recovering from the employer all wage and material increases during that period.

It is inevitable that an employer who has to pay out what may be enormous sums for these will turn round on his architect and say: 'You advised me to use JCT 63 instead of JCT 80. If I had used JCT 80, I would not have to pay the contractor £250,000 extra. You were negligent in the advice you gave me. Compensate me for my loss.'

There are also other good reasons why the ordinary procedure under JCT 63 should not be used.

Extensions of time

The provisions of JCT 80 regarding extension of time for completion of the work are discussed subsequently [7.00 *et seq.*]. But it is worth recording here that, as will be seen in due course, the extension of time provisions are not, as is commonly thought, inserted exclusively in the interests of the contractor. They are essential to protect the interest of the employer so far as recovery of liquidated damages is concerned. Without power to extend time for completion, any delay caused to the contractor by the employer or his agents would put the time for completion 'at large', i.e. not the fixed date specified by the contract but at a 'reasonable time'; and, as a result, the employer would lose his right to liquidated damages: *Percy Bilton Ltd* v. *Greater London Council* (1982).

JCT 80 introduces three new provisions which are not contained in JCT 63. The first is in clause 25.4.8.1 which is a rewrite of the 'Epstein clause' which figures in 23(g) of JCT 63.

It now reads:

> 'the execution of work not forming part of this Contract by the Employer himself or by persons employed or *otherwise engaged by the Employer* . . .' [Author's italics].

As will be seen [7.09] the earlier clause was held, by judicial ingenuity, to be wide enough to cover default by statutory undertakers, such as the electricity, water and gas boards, who had undertaken to the employer to perform more than their statutory duties. But the case in question, *Henry Boot Construction Ltd* v. *Central Lancashire Town Development Corporation* (1980), turned entirely upon a particular arbitrator's finding of facts which the judge was bound to accept.

The new clause removes ambiguity and preserves the employer's right to liquidated damages by making it clear that the architect can extend time.

Clause 25.4.8.2 is equally of value to the employer.

The relevant event is the delay caused to the contractor by 'the supply by the Employer of materials and goods which the Employer has agreed to provide for the Work or the failure to supply'.

Further, by clause 25.4.12, the architect is authorised to extend time if the employer fails to give ingress or egress to and from the site.

Without these new provisions, if any of these relevant events happen, the architect has no power to extend time and therefore the employer loses his right to liquidated damages [7.01].

Furthermore, by clause 23 of JCT 63, that contract does not allow the architect to make an extension of time after the due date for completion of the work.

In the case of what can be termed a 'neutral' event, i.e. one that was not the fault of either contractor or employer, this was unimportant.

The contractor could not expect to be given any extension of time when the real reason for the event was his own delay. If he had not already been in default he would not have suffered further delay, e.g. by exceptionally adverse weather conditions. If the time for completion is November and because of the contractor's failure to complete he is further delayed by heavy snowfalls in February, under no circumstances could he expect an extension of time.

When it did become important was when the event was the fault of the employer or his architect, e.g. late delivery of instructions.

In that case, since the architect had no power to extend time, such events effectively released the contractor from his obligation to pay liquidated damages and time for completion became at large.

As a result, another effect was that the architect could not issue variation orders after the due date for completion had passed, and is applied both to alterations to existing plans and the provision of extra work. Since there was no power to grant an extension of time, the provisions for liquidated damages became invalid.

JCT 80 rectified that situation by empowering extensions of time to be granted even after practical completion, and it also provides for the repayment of liquidated damages upon the architect fixing a later completion date: clauses 25.3.3 and 24.2.2.

Variations can now be made under JCT 80 after the due completion date has passed and the topic is discussed later [7.21].

Comments about JCT 63

The enthusiasm for JCT 63 seems to be a somewhat recent event. It was in *English Industrial Estates* v. *Wimpey* (1972) that Lord Justice Edmund Davies described this contract as 'a farrago of obscurities'. Ian Duncan Wallace QC, the learned editor of *Hudson's Building Law* and *Building and Civil Engineering Standard Forms*, wrote that 'no adviser of any private employer should allow the [1963] forms to be used without substantial amendment'. He suggested even then that any architect who used the 1963 forms could be liable to his employer for negligence.

The 1963 JCT form had major defects and was condemned by judicial and extra-judicial opinion. No architect is therefore justified in using it when there is in existence, and agreed by the whole industry, the JCT 80 form which corrects at least ten of the major defects in the old forms.

1.08 Engagement of nominated and other sub-contractors

In *Westminster* v. *Jarvis* (1970) it was forcefully pointed out by the House of Lords that the JCT 63 contract was inequitable because it allowed the main contractor an extension of time under clause 23(g) for 'delay on the part of a nominated sub-contractor' but allowed the sub-contractor to escape without liability because there was no privity of contract between himself and the employer.

As Lord Dilhorne put it:

> 'It is indeed curious that in this form of contract, issued by the RIBA, and approved by many other bodies, one should find a provision under which a sub-contractor can benefit from his own default.'

And Lord Wilberforce commented:

> 'I cannot believe that the professional body, realising how defective this clause is, will allow it to remain in its present form.'

In response to those criticisms, the RIBA (not the JCT) produced in 1973 the grey 'Form of Warranty', an employer/sub-contractor agreement to be used by sub-contractors invited to tender for nomination.

Unfortunately, this form was often either not used or was signed after the sub-contractor had been nominated, in which case his warranty was without consideration and no more than a worthless piece of paper (at least in England if not in Scotland).

By contrast, the new 'basic' system for nomination of a sub-contractor requires a sub-contractor to enter into contracts NSC/2 or NSC/2a with the employer *before* he gets the nomination. The key to the NSC/2 or 2a contracts is contained in clause 3.4 of NSC/2 and clause 2.4 of NSC/2a. By those clauses the sub-contractor gives a warranty to the employer that:

> 'The sub-contractor shall so perform the sub-contract that the contractor will not become entitled to an extension of time for completion of the Main Contract Works by reason of the Relevant Event in Clause 25.4.7 of the Main Contract Condition.'

In other words, if the employer loses his right to recover liquidated damages because the main contractor has got an extension of time for 'delay on the part of a nominated sub-contractor' then the employer can sue the sub-contractor for the liquidated damages he has lost. (However, there are provisions in NSC/4 and 4a for the sub-contractor also to obtain an extension of time, in which event the employer is still without recourse.)

There are other valuable rights that an employer now has contractually against a nominated sub-contractor under NSC/2 or 2a.

If the architect or his employer prefers not to have any nominated sub-contractors, there is now provision in clause 19.3.1 for the employer to provide the contractor with a list of at least three persons from whom to select a domestic sub-contractor. (Prudently, he will provide five — that is, if there are five.)

Chapter 2

Legal principles

2.01 Legal background to the JCT contracts

'The problem with the Standard Form is that so many people believe that they know precisely what it means, whereas, in practice, hardly anyone does,' wrote an unusually candid quantity surveyor in a trade journal. In part, that is due to the fact that, like all contracts, the Standard Form must be read against a background of law.

The Standard Form must be construed against the background of common law, and in particular those branches of it which deal with contract and torts. Terms are to be implied in it which are not contained in the printed form [2.06]. The obligations of the contractor and the relationships between employer, architect, contractor and sub-contractor will be governed not only by contract but by the law of torts, particularly the torts of negligence and breach of statutory duty.

Moreover, relationships between the parties will also be affected by the principles of equity, particularly since a trust is created in respect of retention monies [6.10]. The assignment of contractual benefits may also be governed by the rules of equity [5.13].

Finally, the words that the contract uses will be affected by express statutory provisions such as those contained in the Limitation Act 1980, the Unfair Contract Terms Act 1977, the Defective Premises Act 1972 and the other pieces of the multitudinous legislation with which Parliament burdens our lives. Of legislation two things can be said: firstly, that most legislation fails to achieve its object; secondly, that all legislation has unanticipated and undesirable side effects. An act is passed to enable injured persons to sue the estate of car drivers who die in a crash (the Proceedings against Estates Act 1970): it results in the estate and beneficiaries of a deceased architect being sued for work he did as a young man.

2.02 Privity of contract

A fundamental principle of English contract law (but not of Scottish law) which affects the Standard Contract is that a person who is not a party to the contract cannot recover a benefit under the contract even though the contract purports expressly to confer that benefit on him. 'Our law knows nothing of a *jus quaesitum tertii* ("a right of a third party") arising by way of contract,' explained Lord Haldane.

The case of *Tweddle* v. *Atkinson* (1861) demonstrates the point. William Guy contracted with John Tweddle that each would pay £200 to William Tweddle, who was the son of one of the contracting parties and son-in-law of the other. The contract purported to give William Tweddle express right to sue either for failure to perform their obligations: 'The said William Tweddle has full power to sue the said parties in any court of law or equity for the aforesaid sum.' William Guy died without having performed his obligations under the contract and William Tweddle therefore sued his executors for the £200. It was held that he could not recover the money since he was a foreigner to the contract.

Conversely, a person who is not a party to a contract cannot have obligations imposed on him by the contract, even if he knows of its terms: *McGruther* v. *Pitcher* (1904); *Adler* v. *Dickson* (1954).

The established law is undoubtedly that a third person who is not a party to the contract cannot take a benefit under it. As Lord Reid said in *Scruttons Ltd.* v. *Midland Silicones Ltd* (1962) AC 466:

> 'I find it impossible to deny the existence of the general rule that a stranger to a contract cannot, in a question with either of the contracting parties, take advantage of the provisions of the contract, even where it is clear from the contract that some provision in it was intended to benefit him.'

In that case, one who was not a party to the contract was not protected by an exemption clause in the contract.

Benefits for sub-contractors

It follows that, since JCT 80 is a contract between the employer and the contractor, even if in that document the contractor promises to confer certain benefits on a sub-contractor, that sub-contractor cannot sue the contractor under JCT 80 for those benefits.

Moreover, although JCT 80 makes frequent references to what the architect shall do or may do, he is not a party to that contract, so that those expressions create no contractual obligation on him to do any of those things. The contractor therefore cannot sue the architect for breach of contract if he does not do anything that JCT 80 says he shall, because there is no contract between the two. Nor can he join the architect in an arbitration arising out of the arbitration clause in the JCT contract.

This cannot be stressed too strongly, since it is frequently overlooked by members of the construction industry: contracts only bind those who are parties to them.

It will be necessary to return to this topic later in connection with the subject of retention of title to goods until payment [6.04]. It has been suggested in the construction press that the industry has nothing to fear from retention of title clauses since clause 14(1) of JCT 63 provided (as clause 16.1 of JCT 80 does in similar terms):

> 'Unfixed materials and goods delivered to, or placed adjacent to the Works, and intended therefore for the Works, shall not be removed except to use upon the Works, unless the Architect has consented in writing. . . . Where the value of any such materials or goods has been included in an interim certificate under which the Contractor has received payment, *such materials and goods shall become the property of the Employer*. . . .'

Commentators overlook two things: firstly, the doctrine of privity of contract; secondly, that a fundamental tenet of English law is expressed in the Latin maxim *nemo dat quod non habet*: 'A man cannot give what he has not got.' That clause is not binding on those not a party to the JCT contract and therefore is not binding on the sub-contractor.

As a general rule, unless the contractor has become the owner of the unfixed materials, he cannot pass a title to the employer [2.03].

Contracting as agents

There are numerous statutory provisions which constitute exceptions to the doctrine of privity of contract such as the situation where the parties contract expressly as trustees in relation to an existing or future money fund or asset. Equity will enforce the trust at the instance of the third party, the beneficiary.

Another apparent exception to the general rule is the law of agency. If one contracting party is expressly authorised to contract on his own behalf and as agent for another who is not on the face of it party to the contract, that third party may rely on provisions of the contract. This is not really an exception to the general principle since the third party has in fact become a party to the contract by his prior authorisation of his agent to make the contract on his behalf.

However, in *New Zealand Shipping Co. Ltd* v. *A. M. Salterthwaite & Co. Ltd* (1975), in exactly the same circumstances as in the *Scrutton* case, an exemption clause in a charter party in favour of stevedores was upheld as the result of the ingenious wording.

This led the draftsmen of the IMechE/IEE Model Form of General Conditions, 1966 edition, to revise clause 30. The contract provides for the issue of a taking-over certificate and then under the heading of 'Defect after Taking-Over' now provides:

> Clause 30 (vii). 'The Contractor's liability under this clause shall be in lieu of any condition or warranty implied by law as to the quality or fitness for any particular purpose of any portion of the Works taken over under Clause 28 and save as in this clause expressed neither the Contractor nor his Sub-contractors, servants or agents shall be liable, whether in contract, tort or otherwise, in respect of defects in or damage to such portion, or for any injury, damage or loss of whatsoever kind attributable to such defects or damage. *For the purposes of this sub-clause the Contractor contracts on his own behalf and on behalf of and as trustee for his Sub-contractors, servants and agents*' [author's italics].

It is noteworthy that this is the contract used in the House of Lords' case of *IBA* v. *EMI and BICC* (1980). An earlier version of clause 30 failed to protect either the main contractor or the nominated sub-contractor in that case. Indeed, although it was relied on at first instance, there is no indication that the clause was relied on in any way in their lordships' house.

Sub-contractors, who were not a party to the main contract between the employer and main contractor in the 'Model' Condition form tried to rely upon it in *Southern Water Authority* v. *Fraversh de Vic Carey and Ors* (1984), which was heard by Mr David Smout QC sitting as a deputy High Court judge.

This argument was rejected by the judge because there was no evidence that the main contractor had ever, *before* the Model Conditions had been entered into, been authorised to contract as the sub-contractor's agent. It was also suggested that, even if the main contractor had not been expressly constituted the sub-contractor's agent at the time when he made the contract, the sub-contractor, as disclosed principal, was entitled to ratify the contract subsequently. This argument was also rejected, the judge saying:

> 'The fact that [the sub-contractors] were in the contemplation of [the main contractors] at the material time, does not in my view suffice.'

To that, there can be no objection. Undisclosed principals cannot ratify: *Maxsted & Co.* v. *Durant* (1901); nor can intended sub-contractors who have not at the date of the contract been nominated.

Can a contract to which a sub-contractor is not a party protect him against a negligence action?

However the judge held, surprisingly, that in spite of the fact that the sub-contractor could not rely upon clause 30 of the Model Conditions since there was no privity of contract between the employer and the sub-contractor, nor could he rely upon it since the main contractor was not in truth his agent, he *could* rely upon it as an indication that the employer had exempted him from any liability for negligence. The present author does not profess to understand this reasoning, so will set out the relevant part of the judgment in full. The judge referred to the 'proximity' test mentioned in the House of Lords in *Junior Books* v. *Veitchi* (1982) and continued:

> 'No one would doubt that in an ordinary building case as between the sub-contractors and the building owner who has suffered damage there is a sufficient relationship of proximity that in the reasonable contemplation of the sub-contractor carelessness on his part may be likely to cause damage to the building owner. Thus a *prima facie* duty of care lies upon the sub-contractor. So also in this case. But one has to go on to consider whether there are any considerations which ought to negative or to reduce or limit the scope of that duty. And merely to ask the question in the context of this case seems to me to foretell the answer. Did not the plaintiffs . . . as building owners, as it were, themselves stipulate that the sub-contractors should have a measure of protection following upon the issue of the appropriate taking-over certificate? We must look to see the nature of such limitation clause to

> consider whether or not it is relevant in defining the scope of the duty in tort.
>
> The contractual setting may not necessarily be overriding but it is relevant in the consideration of the scope of the duty in tort for it indicates the extent of the liability which the plaintiffs' predecessors wished to impose. . . . As the plaintiffs' predecessors did choose to limit the scope of the sub-contractors' liability, I see no reason why such limitation should not be honoured.'

He therefore held that the sub-contractors had no liability to the employer in tort.

2.03 Ownership of building materials

The doctrine of privity of contract and the effect of JCT 63 clause 14 were argued before Mr Justice Mais in the High Court in *Dawber Williamson Roofing Ltd* v. *Humberside County Council* (1979).

Humberside County Council entered into a contract on the JCT 63 form with a main contractor for the erection of a school. Roofing work was sub-contracted on the NFBTE/FASS Blue Form to the plaintiffs. In November 1976, they delivered to the site 16 tons of Welsh roofing slates, which were included in an interim certificate by the architect and paid for by the employers. The main contractors then became insolvent in January 1977 and their employment determined. This, of course, had the effect also of determining the roofing sub-contractors' contract.

The plaintiffs had not been paid for the slates, so they therefore went to collect *their* slates and were met by a refusal on the part of the Humberside County Council to allow them to do so. The Council claimed that, by virtue of clause 14(1) of JCT 63 (now JCT 80 clause 16.1) the property had passed to them.

Dawber Williamson sued for wrongful detention and conversion of the slates.

The judge found in their favour. The plaintiffs, as sub-contractors, were not parties to the JCT contract and therefore could not be bound by anything in it. It was suggested by counsel for the Council that clause 1(1) of the Blue Form was sufficient to incorporate all the terms of the JCT form into the contract between the contractor and the sub-contractor. This clause read:

> 'The Sub-contractor shall be deemed to have knowledge of all the provisions of the main Contract. . . .'

The judge rightly rejected that submission. Knowledge of the terms of a contract does not make a third party a party to that contract. The sole purpose of clause 1(1) of the Blue Form was to saddle the contractor with knowledge of the consequences of any breach by him, so that he could not rely on the principle set forth by Lord Justice Asquith in *Victoria Laundry* v. *Newman Industries* (1949) that:

> 'an aggrieved party is only entitled to recover such part of his loss actually resulting as at the time of the contract is reasonably fore-seeable as liable to result from the breach.'

Deemed to have full knowledge of the main contract, a sub-contractor in breach cannot claim that the liquidated damages which the contractor may have to pay the employer were not 'reasonably foreseeable'.

On the other point, the question of title, it was clear from the Blue Form that the main contractor did not own the 16 tons of Welsh roofing slates and indeed, under the terms of the Blue Form, never would ever have become the owner of them.

That case makes these points clear.

A sub-contractor to the contractor is in no way bound by the terms of the JCT contract between the contractor and the employer. Even had there been an express term in the sub-contract that the sub-contractor should be bound by all the provisions of the JCT contract, that would have been ineffective: since he was not a party to the JCT contract, he could have no obligations imposed on him by it.

Further, under a 'work and materials' contract, it is not intended that the property in goods and materials provided for incorporation in the work shall ever vest in the contractor; it follows that since he has no title, and is never intended to have one, he can have no title to pass to the employer under JCT 63 clause 14(1) or JCT 80 clause 16.1.

It was hard, no doubt, for the Council to have to pay twice for the slates, but no harder than for the sub-contractor to lose his property to them.

This is a particularly valuable illustration of the point that, whatever the parties may say in their written or printed contract, their words may prove entirely ineffective because of some fundamental principle of law which they have ignored.

The implication of this case for architects called on by JCT 80 clause 30 to certify for unfixed materials is discussed later [6.01], as also is the attempt by the JCT to circumvent those principles by the first amendment to JCT 80.

Contracts under seal

The principle of privity of contract may be affected according to whether the contract is under seal or not, since one of the reasons given for the privity of contract doctrine is that no consideration has passed from the third party to the promisors. A contract under seal requires no consideration.

The effect of this on benefits which the contractor, by JCT 80, has undertaken to confer on a sub-contractor has not been adequately explored in construction cases and, so far as the present author can see, there is no authority in English or Commonwealth cases dealing with the point; see also [2.07].

Scottish law and the passing of property

The principles outlined above are deeply rooted in English law. But they are not applicable to Scottish law and it may be that in Scotland clause 16 of JCT 80 would be effective to pass the property to the employer in the case of goods brought on site by a sub-contractor and paid for by the employer.

In Roman Law, on which Scottish law is based, a third party who was the beneficiary of a contract between two other parties could sue to enforce it. Moreover, Scottish law regarding property on the sale of goods is quite different from English law, and the Scottish version of the JCT 80 has been amended accordingly.

Amendment 1 (1984) to JCT 80

The first amendment made to JCT 80 by the Joint Contracts Tribunal was in July 1984 and it was plainly marked 'Not for Use in Scotland', as was the corresponding amendment (No. 2) to NSC/4. Both attempt to reverse the decision in *Dawber Williamson* v. *Humberside County Council* (1979) [2.03].

The JCT amendment is to clause 19.4 and seeks to impose an obligation that the contractor shall include in every domestic sub-contract a clause:

> 'where, in accordance with clause 30.2 of the Main Contract conditions, the value of any such materials or goods shall have been

included in any Interim Certificate under which the amount properly due to the Contractor shall have been discharged by the Employer in favour of the Contractor, such materials or goods shall be and become the property of the Employer and the Sub-contractor shall not deny that such materials or goods are and have become the property of the Employer.'

In other words, if the contractor has been paid by the employer but has not paid the sub-contractor for the goods, the employer shall get a title by estoppel.

For reasons outlined in [2.02], it must still be highly doubtful whether the employer can enforce the term of a contract (i.e. that between the contractor and the sub-contractor) to which he is not a party.

Moreover, if the sub-contractor has no title to the goods, i.e. because his own supplier has retained title to them until payment, he cannot pass one to the contractor. Clearly, the contractor is not a buyer of the goods within the meaning of section 25 of the Sale of Goods Act 1979, since there is no contract of sale of the goods between the contractor and the sub-contractor. So how can he possibly pass a title to the employer?

There are similar provisions in Amendment No. 2 (1984) clause 21.2.4.2 of NSC/4 for nominated sub-contractors.

They are, it is submitted, totally ineffective to pass a title to the employer until the goods are incorporated into the realty.

And why indeed should they be effective? Why should payment by the owner to the contractor for loose goods brought on site deprive a sub-contractor of ownership of his goods, when the contractor may or may not pay him?

The amendment to both JCT 80 clause 19.4.3 and NSC/4 clause 21.4.3 also provides that every domestic and nominated sub-contract shall include the term that, if the main contractor shall have paid the sub-contractor before incorporation in an interim certificate, the property in loose goods on site 'be and become the property of the Main Contractor'.

Domestic sub-contracts are required to include a clause:

'Provided that if the Main Contractor shall pay the Sub-Contractor for any such materials or goods before the value therefore has, in accordance with clause 30.2 of the Main Contract Conditions, been included in any Interim Certificate under which the amount properly due to the Contractor has been discharged by the Employer in favour of the Contractor, such materials or goods shall upon such payment by the Main Contractor be and become the property of the Main Contractor.'

And the NSC amendment reads:

4.3 'Provided also that if the Main Contractor shall pay the Sub-contractor for any such materials or goods before the value therefore has, in accordance with clause 30.2 of the Main Contract Conditions, been included in any Interim Certificate under which the amount properly due to the Contractor has been discharged by the Employer in favour of the Contractor, such materials or goods shall upon such payment by the Main Contractor be and shall become the property of the Main Contractor.'

This is all incredible draftsmanship. If the contractor is paid, the property is supposed to pass to the employer; but when the sub-contractor is paid property is supposed to pass to the contractor.

The construction industry has an invincible conviction that once goods are paid for, ownership passes to the payer, but that is not English law. The industry would have profited from the examination of a case in which the present writer was involved as counsel, before the present Hire Purchase Act came into force, where no less than twelve parties, represented by twelve different counsel, had paid for one motor car and not one, private buyer or dealer, had become the owner.

Even if a sub-contractor is paid for goods, the contractor cannot become the owner of them unless the sub-contractor is the owner, or is in possession of the goods with the consent of the seller on the terms of section 25 of the Sale of Goods Act 1979.

2.04 Work and materials contracts

The *Dawber Williamson* case is illustrative of another point which is often overlooked. Building contracts, which are contracts for work and materials, are not contracts for the sale of goods and are not therefore subject to the Sale of Goods Act 1979, or its predecessor, the Sale of Goods Act 1893.

This has important consequences so far as terms to be implied in a contract are concerned [2.05 *et seq.*], the period in which an action can be brought [2.10] and the passing of title in goods and materials.

Had Dawber Williamson been only the suppliers of goods to the contractors, the contractor might possibly have been able to pass a title by virtue of being a buyer in possession with the consent of the seller under section 25(2) of the Sale of Goods Act 1893 (the statute in force at the time of the transaction) or as it now is, 25(1) of the 1979 Act. But where a contractor employs a sub-contractor to provide work and materials to the employer's property, it is never intended that the contractor should at any time become the owner of materials fixed or unfixed, and he can therefore never be 'a buyer' for the purposes of that section of the Sale of Goods Act 1893, or section 25(1) of the 1979 Act.

The common law provisions which existed before the codification in the Sale of Goods Act 1893, that Act itself, and the current Sale of Goods Act 1979 have no application to these transactions because they are contracts not for the sale of goods but for work and materials.

The distinction was emphasised by the Statute of Frauds 1677, section 17, which later became section 4 of the Sale of Goods Act 1893, and was not repealed until the Law Reform (Enforcement of Contracts) Act 1954 came into force. Under the Statute of Frauds and section 4 of the 1893 Act, contracts for the sale of goods of the value of £10 or more had to be evidenced in writing before they could be enforced: contracts for goods and work were not subject to this provision. To take a modern example, if a lift for a building were ordered then the contract would be unenforceable unless it was in writing; if a lift were ordered to be installed by the makers then the contract would be enforceable even though it was entirely oral.

For nearly 300 years, therefore, there was a sharp distinction drawn between contracts for the sale of goods and contracts for the supply of work and materials, and a vast amount of erudition has been dedicated to elucidating this distinction. Not all of this learning would commend itself to logicians. Why the making and supply of false teeth should be regarded as a sale of goods, but the making and supply of a portrait as a contract for work and materials is not immediately apparent to the simple mind, but that is what has been decided by the courts: *Lee* v. *Griffin* (1861); *Robinson* v. *Graves* (1935).

Nor is it immediately apparent why the serving of a meal in a restaurant is the sale of goods but putting a hair dye on a customer is work and materials: *Lockett* v. *Charles* (1938); *Watson* v. *Buckley, Osborne, Garrett & Co. Ltd* (1940).

Formerly it was said that when 'the contract is such that a chattel is ultimately to be delivered . . . the cause of action is goods sold and delivered': *Lee* v. *Griffin* (1861). But that is no longer a reliable test, and there can be little doubt that if the contract requires the design, fabrication and installation in a building of an air-conditioning plant, for example, it will be a contract for work and materials.

It is not always easy to distinguish between the two in law, as the cases mentioned earlier make clear. But it may be vital. In Australia it has been held that a contract for the supply and installation of a revolving cocktail cabinet which slotted into fastenings on the floor and ceiling was a contract for work and materials: *Brooks Robinson Pty Ltd* v. *Rothfield* (1951). So also were seats installed in a lecture theatre: *Aristoc Industries* v. *R.A. Wenham (Builders) Pty Ltd* (1956); and a lift installed in a building: *Sydney Hydraulic & General Engineering* v. *Blackwood* (1908). On the other hand the installation of a domestic heater was held to be a contract of sale: *Collins Trading* v. *Maher* (1969).

In England, it has been held that where the method of installation is a minor part of the work the contract is for the sale of goods, so that fitted carpets come in this category: *Philip Head & Co.* v. *Shopfronts Ltd* (1970).

2.05 Implied terms on the sale of materials

Under the Sale of Goods Act 1979, a condition is implied in all sales of chattels, that the seller has a right to sell the goods, that the buyer shall enjoy quiet possession of them and that they are free from any charge or incumbrance to a third party (section 12).

There are also implied terms as to quality of the goods: if the goods are sold by description, that they will correspond with the description (section 13); that the goods are of 'merchantable quality' (section 14); and that the goods are reasonably fit for the purpose required (section 14).

These implied warranties could formerly be excluded by an express contractual term, and frequently were for many years, but by the Supply of Goods (Implied Terms) Act 1973 (since replaced) this, in most cases, became impossible. The position, so far as goods are concerned, is now governed by section 6(2) of the Unfair Contract Term Act 1977, which provides that against a person dealing as a consumer (which is defined), liability for breach of these implied warranties cannot be excluded or restricted by reference to any contract term. As against a person not dealing as a consumer, section 6(3) provides that the warranties can only be excluded or restricted by reference to a contract term only so far as the term is reasonable.

This fully applies to goods supplied to a building site for incorporation into the structure.

At a later stage in this book [5.17] it will be necessary to consider JCT 80 clause 36 on nominated suppliers in the light of the Sale of Goods Act 1979, and see whether a contract with a nominated supplier which follows the requirements of clauses 36.4 and 36 is affected by that statute.

It is sufficient for the moment to point out that these terms are implied by statute, irrespective of what the contract may say, and that they are not easily avoided.

The implied warranties as to quality are quite irrespective of fault on the part of the seller. He will be liable even if he does not know of any defect and even if he has exercised every possible care. In effect, he is made to guarantee the goods he sells.

These are the terms implied by operation of law — in this case, statute law. The Supply of Goods and Services Act 1982 also imposes similar implied warranties on certain contracts other than the sale of goods where title is intended to pass under the contract.

2.06 Types of implied terms

Lord Wright said in *Luxor (Eastbourne) Ltd* v. *Cooper* (1941) that there were three types of implied terms. In fact, at least eight different types can be distinguished.

These are terms which the courts will write into contracts even where the parties have not included them. It is necessary, therefore, with all contracts to consider not only the express terms of the contract (i.e. the ones that the parties have in fact agreed or which they will be taken to have agreed) but also the terms that a court may imply into the contract.

Implied terms may be included by:

(a) Local custom: *Brown* v. *I.R.C.* (1965).
(b) The usage of a particular trade, e.g. a bakers' dozen being thirteen or, in *Symonds* v. *Lloyd* (1859), 'reduced brickwork' meaning brickwork 9 inches thick.

(c) The parties' 'own dictionary' usage — where both have agreed on using language with a meaning different to the common one: *The Karen Oltmann* (1976).
(d) As a basis for the doctrine of frustration of contracts where there is a supervening event which prevents performance: *Davis Contractors Ltd* v. *Fareham Urban District Council* (1956). Although this theory is no longer fashionable, it is used in this sense in many reported cases.
(e) Imposed by statute, as has been seen in the Sale of Goods Act 1979, the Supply of Goods and Services Act 1982 and the Defective Premises Act 1972.
(f) By common law [4.02 *et seq.*].
(g) To give commercial effectiveness to a contract — the doctrine of *The Moorcock* (1889).

A jetty owner entered into a contract with the owner of the ship called *The Moorcock* to discharge her at the berth alongside their jetty. Both parties knew that when the tide ebbed the ship would take the ground. When she did, she suffered damage as the result of settling on a hard ridge. There was no express promise in the contract that the berth was safe for the ship, but the court held that one would be implied.

> '*Prima facie* that which in any contract is left to be implied and need not be expressed is something so obvious that it goes without saying; so that if while the partiers were making their bargain an *officious bystander* were to suggest some express provision for it in their agreement, they would testily suppress him with a common "Oh, of course"': MacKinnon LJ.

From this it is clear that there can never be an implied term of *this nature* if there is an express term dealing with the same matter: *Les Affréteurs Réunis* v. *Leopold Walford* (1919). But it is sometimes erroneously supposed that this principle applies to all implied terms. It does not apply to those terms which are to be implied by law i.e. classes (e) and (f) above.

The question of what terms the court will imply into the JCT 80 contract and of what kind is discussed later under the contractor's obligations [4.01 *et seq.*] and in relation to:

— The Building Regulations [4.09]
— Extensions of time [7.04].

2.07 Simple contracts and contracts under seal

The Standard Form of Building Contract can be entered into in two forms: as a simple contract or as a specialty contract, that is under seal — what is called 'a deed'.

All the great cathedrals and buildings in England were erected under obligations assumed by means of deeds. Many of these documents are still extant and a typical example, dated 1239 for the construction of a wharf on the River Thames begins: *Haec indentura facta inter Decanum et Capitalum ecclesie Santi Pauli London ex parte et Richardum Coterel carpentarium ex parte altera testatur. . . .* It would be very similar to a deed in use today: 'THIS INDENTURE made between the Dean and Chapter of St Paul's Church, London, of the one part and Richard Coterel, carpenter, of the other part WITNESSETH. . . .'

Originally only obligations contained in such documents under seal were enforceable in the king's courts (as distinct from local courts or merchants' courts), but in the sixteenth century contractual promises made orally or in writing became enforceable by the action of *assumpsit*, which was in essence a tort action. In fact, it was not until the first quarter of the nineteenth century that the law of contract disentangled itself from the 'action on the case' for the tort of *assumpsit.* Within the last decade, contract and tort appear to be converging again, with the substantial difference that contractual obligations are those which are voluntarily assumed by the parties whereas the obligations in tort are those imposed by law on everybody.

Many building contracts in the Standard Form are still executed under the seal of the parties: that is, in the form of deeds. English law has a superstitious reverence for such pieces of paper although, as will be seen, they differ little nowadays, except in their effect, from the ordinary contract in writing.

Contracts under seal are termed 'specialty contracts' and differ from ordinary contracts in writing in three respects:

(1) Consideration is not necessary to support promises made under seal. That is, a gratuitous promise to do something will be enforceable even though the promisee has neither given nor promised to give anything in return. It might therefore be that a promise to confer a benefit on a sub-contractor, contained in a

JCT 80 contract under seal between the employer and the contractor, might well be enforceable by the sub-contractor, notwithstanding the doctrine of privity of contract [2.02].

(2) The parties are not in theory permitted to deny statements of fact contained in deeds, including the recitals of the deed. The parties are, as the lawyers say, 'estopped by deed'. Thus, there may be a material difference in the interpretation of a JCT contract under seal, and one simply under signature: *Greer* v. *Kettle* (1938).

The most important practical difference is that:

(3) Under the Limitation Act 1980, an action for breach of a contract made orally or in writing ('a simple contract' as these are termed) will be statute barred six years from the date when the 'cause of action accrues'; whereas in a contract under seal, the period is twelve years.

For many years up to 29 July 1960, contracts with corporate bodies had to be under seal. It is no longer necessary but, as a matter of tradition, many local authorities still execute contracts under seal. The effective reason for doing so is to have an extended period in which to bring actions for breach of contract [2.10].

What is a deed? Seals were used, when few could write, to signify a man's assent to a document. Indeed, until 1925, deeds did not have to be signed. They were regarded as so important that the Lord Keeper of the Great Seal of England (which office later became merged with that of the Lord Chancellor) always slept with it under his pillow at night. To forge it carried the death penalty — in spite of that, forgeries were on sale in Fleet Street in the reign of James I for seven shillings.

Nowadays, it is no longer the practice to actually use sealing wax and an impression by a seal. In *First National Securities* v. *Jones and Ano.* (1975), a County Court judge refused to recognise as a deed a legal charge where the mortgagor had simply signed a bank's printed form on which there was a printed circle enclosing the letters 'L.S.' (meaning in Latin 'the place for the seal'). The Court of Appeal set aside this decision. Lord Justice Buckley said: 'The mortgagor has placed his signature across the circle. That is sufficient evidence that the charge was executed as the mortgagor's deed.'

As long as a document is expressed as being a deed and is duly signed as such, it is a deed even if it bears no seal, wafer or impression.

2.08 Liability in contract and tort

For many years, it was an accepted principle of English law that where there was a contract in existence between two parties, the obligations between those two were to be found only in the contract and neither was under a duty to the other apart from the contract. The only exceptions to this principle concerned persons who carried on certain callings such as that of common carrier, railways and innkeeper who, it was said, had dual liability in contract and in tort because of their status.

That view was confirmed as recently as 1964, by the case of *Bagot* v. *Stevens Scanlan* in which Lord Justice Diplock, as he then was, sitting as a judge of the High Court, decided in a closely reasoned and logical judgment, that an architect owed no duty to his client in tort for negligence, but that his duties arose solely and exclusively under his contract with his client.

That decision was criticised by Lord Denning *obiter* in *Esso* v. *Mardon* (1976), on the basis of a long forgotten House of Lords' decision of 1844, and eventually in *Midland Bank Trust Co.* v. *Hett, Stubbs and Kemp* (1979) Mr Justice Oliver declined to follow it and the other authorities, and held that solicitors owed a dual liability to their clients, in contract and in tort. This was approved by the Court of Appeal (Lord Denning presiding) in *Batty* v. *Metropolitan Properties Ltd and Ors* (1979) and followed in *Ross* v. *Caunters* (1980).

The present position therefore is that all contracting parties owe a duty to the other party not only in contract but also in tort.

This means that, quite irrespective of any breach of terms of the JCT 80 contract, a contractor may be liable to his employer for workmanship which is negligent. In the *Batty* case, referred to above, the Court of Appeal in effect equated negligence with breach of the implied warranties now imposed on contractors [4.03]. Failure to investigate sufficiently the terrain on which the house was built was treated as negligence on the part of both the developer and builder and amounted to the same thing in the circumstances of that case as breach of the developer's express warranty that the house would be reasonably fit for use as a dwelling house.

It is possible, subject to the provisions of the Unfair Contract Terms Act 1977, for a contractor to exclude or limit his liability to the employer for the tort of negligence but the JCT 80 contract does not do so expressly, nor does it exclude any implied terms. Dual liability in contract and in tort to an employer has most serious consequences so far as limitation of liability is concerned.

Whether a contractor's or sub-contractor's liability to second and subsequent owners or occupiers can be reduced or eliminated by the terms of the contract with the original employer seems doubtful, although it has been so held by a deputy High Court judge: *Southern Water Authority* v. *Fraversh de Vic Carey and Ors* (1984) — see [2.02].

As a matter of principle, since the contractor has no contractual relationship with subsequent owners or occupiers and in no way can they have exonerated the contractor from the duty of care imposed by law on everybody, it is difficult to see how the terms of the contractor's contract with his employer can in any possible way affect the duty of care he owes to third parties.

2.09 Breach of statutory duty

Apart from liability to his client for the tort of negligence the contractor may be liable to him also for the tort of breach of statutory duty; and also to third parties.

At the time when the first edition of this book was written there were some 466 Acts of Parliament on the Statute Books and numerous Statutory Instruments which constituted some 7208 separate criminal offences. There are, inevitably, more now.

Apart from creating criminal offences, much of this legislation enables a person who has been injured or suffered loss to sue the offender (criminal is too harsh a word, though entirely applicable) for damages — quite apart, of course, from any penalty imposed on him for the breach.

Not every statute creates tortious liability. Some expressly exclude it. The Medicines Act 1968 says: 'The provisions of this Act shall not be construed as conferring a right of action in any civil proceedings.'

Some expressly confer it. Most are silent about whether or not there is a civil liability, as well as criminal, for breach of the statute. In *Monk* v. *Warbey* (1935) Lord Justice Greer said:

> '*Prima facie* a person who is injured by breach of a statute has a right to recover damages from the person committing it, unless it can be established by considering the whole of the act that no such right was intended to be given.'

Even where an act is passed for the protection of the public in general, a plaintiff can sue for breach of statutory duty over and above that suffered by the rest of the public: *Couch* v. *Steel* (1854). The Law Commission, in its report No. 21 (1969), recommended the enactment of a statute that would provide:

> 'Where any act . . . imposes or authorises the imposition of a duty whether positive or negative . . . it shall be presumed, unless express provision to the contrary is made, that a breach of duty is intended to be actionable . . . at the suit of any person who sustains damage in consequence of the breach.'

Guidance as to what statutes constitute those for which an action for damages will lie can be gathered from the Court of Appeal decisions in *Monk* v. *Worby* (1935), *Square* v. *Model Farm* (1939), *McCall* v. *Abelesz* (1976) and the House of Lords' decision in *Cutler* v. *Wandsworth Stadium* (1949).

There has been some controversy as to whether the Building Regulations constitute a statute for breach of which an action for damages will lie. As long ago as 1954 Lord Goddard LCJ in *Solomons* v. *Gertzenstein* held that an action would lie for breach of the London Building Acts. There appears to be no distinction in principle between those statutes which apply in the metropolis and those which apply in the rest of the country.

Lord Wilberforce in *Anns* v. *London Borough of Merton* (1978) certainly thought that an action could lie. He said:

> 'I agree with the majority in the Court of Appeal in thinking that it could be unreasonable to impose liability in respect of defective foundations in the council, if the builder, whose primary fault it was should be immune from liability . . .'

He concluded:

> 'Since it is the duty of the builder (owner or not) to comply with the by-laws . . . an action could be brought against him . . . for

breach of statutory duty by any person for whose benefit or protection the by-law was made.'

That was not a piece of *obiter dicta*; it was clearly part of the rationale by which he arrived at his final decision. Even if it had been purely *obiter dicta*, those expressed in opinions in the House of Lords, especially those which carry, as this did, the assent of a majority of law lords, are usually regarded as binding on all inferior courts.

This was followed by the Official Referees in several cases, two of which are *Eames London Estates Ltd* v. *North Herts District Council* (1980) and *Acrecrest Ltd* v. *W. S. Hattrell & Partners Ltd* (1980).

Unhappily, Mr Justice Woolf in *Worlock* v. *Saws* (1981) without hearing any argument on the point from counsel, in an *obiter dictum* dissented from this view. Further, in another *obiter dictum* and again without argument, another judge, Lord Justice Waller, 'without expressing a concluded opinion' in *Taylor Woodrow* v. *Charcon and Ors* (1982) supported this view.

Unfortunately, Judge John Newey QC in *Perry* v. *Tendring District Council* (1984) followed these two *obiter dicta* and said:

'In my opinion, building by-laws do not give rise to liability in damages but I trust before long there will be a specific decision of the Court of Appeal on the subject.'

None of these three judges had his attention drawn to section 71(1) of the Health and Safety at Work Act etc. 1974 which provides that 'a breach of duty imposed by the building regulations shall, so far as it causes damage, be actionable except so far as the regulations otherwise provide'.

The regulations do not otherwise provide.

2.10 Limitation periods for breach of contract

The Limitation Act 1980, a consolidating statute like its predecessors, specifies a limitation of six years for actions based on simple contract or tort.

The English Limitation Act does not extinguish the right to sue: it merely limits the period within which any particular plaintiff must commence his action if he is not to be barred by elapse of time;

'statute barred' as the lawyers call it. Since the Act does not extinguish a right of action, unless the defendant raises the plea in his defence that the action is statute barred by the Limitation Act, the plaintiff can proceed, as was the case in *McGuirk* v. *Homes (Basildon) and Ano.* (1982) where the contractors were held liable for work done twenty years before.

This is not the situation in Scotland. There, the right to sue is entirely extinguished by 'negative prescription', as in most countries where Roman law principles prevail. By the Prescription and Limitation (Scotland) Act 1973, as amended in 1980, unless a plaintiff brings his action for breach of contract, tort or breach of statutory duty within five years he loses all rights. But the five year period dates only from the time when he first knew or could with reasonable diligence have discovered that he had suffered the loss or sustained the damage.

In England, the six year period (twelve years in the case of contracts under seal) begins to run 'when the cause of action accrues'. In the case of breach of contract, it has been held that the cause of action accrues when a breach of contract took place, whether the plaintiff knew of it or not.

Generally speaking, therefore, a contractor can rest assured that six years after he has completed any particular work no action can be brought against him for breach of contract, unless the contract is under seal [2.07], when the period will be twelve years.

It is clear therefore that, in view of the uncertain state of the law regarding limitation periods in tort action [2.13], there are advantages for every employer to require his main contractor to enter into a contract under seal. And if the main contractor does so, he should ensure that every sub-contractor whether nominated or domestic, does the same.

There are, however, two exceptions to this: the provisions of section 32 of the Limitation Act 1980, and where the contractor has given indemnities to the employer.

2.11 'Fraud' of a contractor

Section 32 of the Limitation Act 1980 provides that time shall not start to run where the right of action has been concealed by the fraud of the defendant. Normally in law 'fraud' involves moral obliquity, a deliberate intent to cheat.

But in a succession of cases relating to building contracts, the courts have placed a meaning on the word 'fraud' entirely inconsistent with its natural or normal meaning. It has been held to mean no more than that it would be inequitable to allow the defendant to rely on the statutory defence provided by Parliament.

The doctrine started with *Beaman* v. *A.R.T.S. Ltd* (1949), continued with *Clark* v. *Woor* (1965) (in which case the builder had substituted Ockley facing bricks, many of which were seconds, for best Dorkings); and received the enthusiastic approval of the Court of Appeal in a case of defective foundations: *Applegate* v. *Moss* (1971).

As a result even in cases of breach of contract, the cause of action does not accrue, where the builder has concealed defects, until the plaintiff knows of the defects.

A plaintiff can rely upon this section even if he himself employs an architect or surveyor to supervise the construction: *London Borough of Lewisham* v. *Leslie* (1979), a case where ties to cladding were omitted or were defective.

However, the concealment must, it seems, be deliberate. In the case of *Street* v. *Sibbabridge* (1980), where the builder built, in accordance with the design, strip foundations 2 ft 9 in deep and did not appreciate that these were inadequate for lias clay soil with a row of trees in the vicinity, it was held that he could rely upon the Limitation Act, since the case was pleaded only as a breach of contract and the breach occurred more than six years before the writ was taken out. Judge Fay declined to apply section 26 of the Limitation Act 1939 and said: 'There must be some degree of moral obliquity; not deliberate, conscious fraud but something other than straightforward behaviour judged by a subjective standard.'

Referring to Lord Denning's judgment in the *Applegate* v. *Moss* case, he said:

> 'I think it is implicit that the builder knows he has done bad work. . . . That is certainly applicable to a case where a builder has skimped on the foundations, or used weak cement or something of that kind. He would know he was taking a chance and for that reason, an element of unconscionableness attaches to his physically concealing the foundations. Lord Justice Edmund Davies in that case said: "It is a truism that not every breach of contract arising from a defect in the quality of materials or workmanship would justify a finding of fraud".'

Later, after discussing the case of *King* v. *Victor Parsons & Co.* (1973), he concluded:

> 'It is necessary to look into that individual's mind and see whether or not he was aware that he was doing wrong or something which might be wrong. If he was aware that he was taking a risk, he was behaving unconscionably. If, however, by an honest blunder he unwittingly commits a wrong, he can avail himself of the Act because he is not acting unconscionably': *Street* v. *Sibbabridge Ltd* (1980).

Judge Fay did not have to consider how this subjective test can be applied to a large company with many thousands of employees; in the case he was dealing with, it was a small limited company with two directors, one only of whom was concerned with the work, and no more than twenty employees.

The relevant section of the 1980 Act reads:

> 32(1) 'where in the case of any action for which a period of limitation is prescribed by this Act, either —
>
> (a) the action is based upon the fraud of the defendant; or
> (b) any fact relevant to the plaintiff's right of action has been deliberately concealed from him by the defendant; or
> (c) the action is for relief from the consequences of a mistake;
> the period of limitation shall not begin to run until the plaintiff has discovered the fraud, concealment or mistake (as the case may be) or could with reasonable diligence have discovered it.
>
> References in this subsection to the defendant include references to the defendant's agent and to any person through whom the defendant claims and his agent.
>
> (2) For the purposes of subsection (1) above, deliberate commission of a breach of duty in circumstances in which it is unlikely to be discovered for some time amounts to deliberate concealment of the facts involved in that breach of duty.'

Section 32 of the 1980 Act is a revision of Section 26 of the 1939 Act made on the recommendations of the Law Reform Committee in their 21st Report (1977) but it is doubtful whether it substantially alters the law. Cases under the former section 26 made the position clear. 'Breach of duty' in section 32(2) includes, of course, negligence — breach of a duty to take care and breach of statutory duty [2.09].

2.12 Limitation periods and indemnities

An indemnity is a contractual obligation by one party to reimburse another against loss. A typical example is to be found in the contractor/sub-contractor contract form NSC/4.

> 5.1.2 'The Sub-Contractor shall indemnify and save harmless the Contractor against and from:
> .2.1 any breach, non-observance or non-performance by the Sub-Contractor or his servants or agents of any of the provisions of the Main Contract referred to in clause 5.5.1; and
> .2.2 any act or omission of the Sub-Contractor or his servants or agents which involves the Contractor in any liability to the Employer under the provisions of the Main Contract referred to in clause 5.1.1.'

This replaces clause 3 of the old NFBTE/FASS Green Form which read:

> 'The Sub-Contractor shall:
> (a) Observe, perform and comply with all the provisions of the Main Contract on the part of the Contractor to be observed, performed and complied with so far as they relate and apply to the Sub-Contract Works (or any portion of the same) and are not repugnant or inconsistent with the express provisions of this Sub-Contract as if all the same were severally set out herein; and
>
> (b) Indemnify and save harmless the Contractor against and from:
> (i) any breach, non-observance or non-performance by the Sub-Contractor, his servants or agents of the said provisions of the Main Contract or any of them; and
> (ii) any act or omission of the Sub-Contractor, his servants or agents, which involves the Contractor in any liability to the Employer under the Main Contract; and
> (iii) any claim, damage, loss or expense due to or resulting from any negligence or breach of duty on the part of the Sub-Contractor, his servants or agents (including any wrongful use by him or them of the scaffolding referred

to in Clause 18 or this Sub-Contract or other property belonging to or provided by the Contractor); and

(iv) any loss or damage resulting from any claim under any statute in force for the time being by an employee of the Sub-Contractor in respect of personal injury arising out of or in the course of his employment.'

I. N. Duncan Wallace in his *Building and Civil Engineering Standard Forms* (1969) at page 195, comments:

> 'Sub-paragraph (i) is strictly unnecessary, since damages amounting to an indemnity would follow from a breach of paragraph (a) and the words the "said provisions" in this sub-paragraph clearly indicate the provisions referred to in paragraph (a).'

With respect to the learned author, there is a fundamental difference between the obligations under 3(a) and 3(b)(i) which, so far as the contractor is concerned, makes the latter sub-clause far from unnecessary.

The same applies with equal force to the corresponding clauses of NSC/4 clause 5.1.1 and clause 5.1.2.1.

In the case of clauses 3(a) of the Green Form and clause 5.1.1 of NSC/4, if there is a breach of this obligation it is statute barred as a breach of contract six years from the date when the work was done by the sub-contractor.

In the case of clauses 3(b)(i) of the Green Form and clause 5.1.2.1 of NSC/4, *the six year limitation period only begins to run from the time when the contractor's liability to the employer has been established and ascertained.*

A sub-contractor does his work and withdraws from the site. Six years later he can know with certainty that, so far as his contractual liability under clause 3(a) of the Green Form and clause 5.1.1 of NSC/4 are concerned, he has no further liability, unless he can be caught by section 32 of the Limitation Act 1980 [2.11].

But under the other two clauses, he may do his work and retire from the site long after practical completion of the main work, and after the almost inevitable arbitration or litigation between the main contractor and the employer has been concluded, the contractor has six years further in which to sue the sub-contractor. It is quite possible in fact for the sub-contractor to be subject to a quarter of a century of possible liability under these clauses.

The general rule is that a person seeking to enforce an indemnity can do so only after the fact and the extent of his own liability has been determined and ascertained.

In a case on the Green Form, Mr Justice Swanwick held in *County and District Properties Ltd* v. *Jenner & Sons Ltd* (1976) that the sub-contractors were not protected against an action under clause 3(b)(i) even though more than six years had elapsed since they did the work, and that time only began to run when the liability of the main contractor had been determined. In the case of the nominated supplier of windows, however, he held that this transaction was subject to the Sale of Goods Act 1893 and therefore any action against the supplier was statute barred. He refused to accept the argument that a similar indemnity to that contained in the Green Form should be implied in the case of a nominated supplier.

That decision was approved by Mr Justice Dillon in *R. H. Green and Silley Weir Ltd* v. *British Railways Board* (1980), who pointed out that it was consistent with the Court of Appeal decision in *The Post Office* v. *Norwich Union Fire Insurance Ltd* (1967), an insurance case. In the *Green* case, a reclamation and dredging company had filled in a railway embankment, the effect of which was alleged to have been that support was withdrawn from the plaintiff's land. The contractor had given the Board an express indemnity against any liability whatsoever as a condition of being allowed to fill in the embankment. When the Board was sued by the plaintiffs, it sought to bring in the contractor as a third party under the indemnity and was met by the plea that such proceedings were statute barred under section 2(1) of the Limitation Act 1939. It was claimed that the Board came under liability, if at all, at the time when the work was done. Mr Justice Dillon rejected this view, in spite of support for it to be found in the case of *Bosma* v. *Larsen* (1966), and said:

> 'Following the reasoning in *The Post Office* v. *Norwich Union* and of Mr Justice Swanwick [in *County and District Properties*], I hold that time does not run against the British Railways Board in favour of [the proposed third party] until the Board's liability to Green has been established and ascertained.'

It follows that the converse must be true: the sub-contractor is under no liability under clause 3(b) of the Green Form and clause 5.1.2 of NSC/4 until any liability to the employer has been established and ascertained. Indeed, that is what the case of *The Post Office* v. *Norwich Union Fire Insurance* expressly decides.

From this it must follow that the contractor can have no right of set-off in equity against the subcontractor under NSC/4 clause 5.1.2 until his, the contractor's, liability to the employer has been determined by agreement, an award or a judgment. But, of course, he may have a right to raise a claim for unliquidated damages for breach of NSC/4 clause 5.1.1.

In the *Country and District Properties* case, *supra*, Mr Justice Swanwick did, however, allow the sub-contractors to be joined as third parties even though there could be no liability on them unless, and until, he found the main contractor liable to the employer.

Indemnities given by the contractor to the employer under JCT 80 are discussed subsequently [4.16].

2.13 Limitation periods in tort

Many actions in tort, whether for negligence, breach of statutory duty or other recognised civil wrongs, are historically based on the old actions of 'trespass on the case'. The simple action of trespass to land was always actionable *per se*, that is, without proof of damage, and so it is to this day. If a contractor swings a crane across your garden, without doing any damage whatsoever, you can bring an action against him for trespass and recover damages — if only nominal ones. Your rights have been infringed.

But in the later 'trespass actions on the case', it was not only necessary for the plaintiff to prove that a wrongful act (or omission) had taken place, but also that he had suffered damage, or loss, as a result. The plaintiff therefore could not have a cause of action until he had suffered damage or incurred loss.

The Limitation Act 1980 provides the same period — six years — for an action to be brought in tort as for a simple contract [2.07]. The time begins to run when the 'cause of action accrues'. When is that?

Where the tort is trespass (or any of the other torts which are actionable *per se*, without proof of damage) clearly the cause of action accrues when the wrongful act is committed.

But in all other torts, which were originally 'actions on the case', three incidents can be distinguished:

(a) the date of the wrongful act or omission

(b) the damage which resulted from that act or omission, which may be:
   (i) physical damage to the plaintiff or his property — whether that be goods or buildings
   (ii) if physical, the first or initial occurrence, followed possibly by subsequent progressive and increasing damage
   (iii) damage in the sense of financial loss (or injury to reputation)
   (iv) latent damage, i.e. that which has as yet not manifested itself but will inevitably do so in the future, as in the case of where it is necessary to execute repairs to avoid injury to health or safety.

(c) the time when the plaintiff first discovers he has suffered damage or loss.

Sometimes, the three incidents can be contemporaneous: a man receives a punch on the nose. That is the wrongful act; that is the time he suffers physical damage; that is the time he knows about it.

In other circumstances, vast periods of time may separate the three events. An employer, in breach of his statutory duty, negligently fails to provide adequate filtration in a factory where men work in clouds of dust: that is his wrongful act or omission. Fifteen years later, that causes a disease in a man's lungs: that is the damage. But it may be another ten years before that is accurately diagnosed and he knows about it.

When does the 'cause of action' accrue?

Counsel for the defendants in the famous case of *Dutton* v. *Bognor Regis* (1972), which established the liability of local authority employers for negligent supervision by building inspectors, complained that if his clients were held liable, under the Limitation Act 1939, their potential liability would extend to infinity. Not so, he was assured by Lord Denning MR in his judgment. The limitation period began to run when the negligence approval was granted.

In other words, the cause of action arose when the negligent act or omission took place, i.e. incident (a) above.

However, less than four years later, Lord Denning, having considered the matter and consulted his brethren, resiled from that opinion in *Sparham-Souter* v. *Town and Country Developments* (1976). He then said:

> 'In recent years the law of negligence has been transformed out of all recognition. The present case is the first one in which this Court has had to consider the effect of that transformation on the statutes of limitation. *One thing is quite clear. A cause of action for negligence accrues not at the date of the negligent act, but at the date when damage is sustained by the plaintiff. . . .*
>
> A builder negligently makes foundations. The Council's inspector negligently passes them as sufficient.
>
> The house is built and sold to a purchaser, who sells it to another and so on down a chain. Some time later the house begins to sink and cracks appear in the structure owing to the insufficient foundations.
>
> The man who is the owner at that time has a cause of action against the builder. And also, under *Dutton* v. *Bognor Regis*, against the Council.
>
> It accrues not at the time of the negligent making or passing of the foundations nor at the time when the latest owner bought the house. It accrues at the time when it begins to sink and the cracks appear. This is the first time any damage is sustained.
>
> It may seem hard on the builder or council surveyor to find himself sued after many years, but it would be harder still on the householder that he should be without remedy, if the surveyor had passed the bad work and the builder covered it up and this prevented it from being earlier discovered' [author's italics].

In that passage, he appeared to be opting for the cause of action accruing when the damage took place, i.e. incident (b) above. But he then went on to say:

> 'The cause of action does not accrue and time does not begin to run until such time as the plaintiff discovers it has done damage or ought, with reasonable diligence, to have discovered it.'

That is to say incident (c) above. This became known as the 'discoverability test'.

Lord Denning's judgment demonstrated regrettably a certain lack of logic. For he assumed that the time when damage occurred was the time when the plaintiff discovered it had happened or could, with reasonable diligence, have discovered it. And he left entirely undefined what 'damage' meant.

The other two judgments, those of Lord Justice Roskill, as he then was, and Lord Justice Geoffrey Lane, were more reasoned.

Lord Justice Roskill rejected the suggestion that when a subsequent owner bought the house or flat, a cause of action 'came with the property'. 'This, with respect is an impossible argument,' he said and gave logical reasons why. He referred to *Halsbury's Laws of England*, Volume 20, page 618, in which it was said, with judicial approval, 'a cause of action cannot accrue unless there is a person in existence capable of suing and another person in existence capable of being sued'. He asked:

> 'How can the statute run against a person who had no title at the time of the alleged negligent act, had suffered no damage and thus at all material times had no enforceable cause of action?'

He concluded that there were two prerequisites to a cause of action: when the plaintiff acquired an interest in the property and when 'the defective state of the property first appears'. In particular, it was pointed out by Lord Justice Geoffrey Lane that a plaintiff who was a second or subsequent purchaser could not possibly have a cause of action until he acquired title in the property.

> 'I am unable to see how it can be said that a cause of action accrues to the plaintiff, and time starts to run against him, before he has obtained any interest in the object . . . be it a chattel or a house.
>
> There is no proper analogy between this situation and the type of situation exemplified in *Cartledge* v. *Jopling*. . . .
>
> The cause of action accrues when the damage caused by the negligent act is suffered by the plaintiff and that cannot be before that damage is first detected or could by the exercise of reasonable skill or diligence have been detected.'

As a result of that decision the principle that the cause of action accrued on the 'discoverability test' became accepted law, especially since the court certified that it was a proper case for appeal to the House of Lords. There was no appeal.

The principle of 'discoverability' was adopted, as it had to be, by High Court judges and the Official Referees. This proposition appears also to have been accepted by the judiciary in Australia and Canada.

It accorded with the basic concepts of the justice. Why should a plaintiff be deprived of a right to sue *before* he knew he had such a right? But it did not, of course, please architects and civil engineers, still less local authorities when they were sued as employers of negligent building inspectors.

However, the test of 'discoverability' was expressly approved by the House of Lords in *Anns* v. *Merton London Borough Council* (1977). That is not the view of the present author alone, but also of Judge Fay QC in *Eames and London Estates and Ors* v. *North Hertfordshire District Council and Ors* (1980), and by other judges in numerous other cases that it would be tedious to mention. It is also the view of Mr Ian Duncan Wallace QC, the learned editor of *Hudson's Building and Engineering Contracts*, who in the first issue of *Construction Law Journal* (1985) wrote:

> 'In *Anns*, the House of Lords appeared to hold quite clearly that time could not run against the plaintiff until the defect in the building *could with reasonable diligence have been discovered by its current owner*: see *Dennis* v. *Charnwood Borough Council* (1982) CA — [original author's italics].

Let us start with the facts of the *Anns* case. The block of flats in Wimbledon was constructed, it subsequently turned out (though no reference to this is to be found in the judgments), partly on the site of a previous house where there was a substantial cellar, over which was built a corner of the new block. This cellar had been infilled with uncompacted rubble, across which ran strip foundations which, it was conceded, had been inspected and approved by the council's building surveyor. He had subsequently moved to Australia and could not be traced to give evidence.

The action was brought when the inevitable happened — the whole block began to tilt with consequential damage to the whole structure. Some of the plaintiffs were the original purchasers of 99 year leases, and others were second or subsequent owners of the unexpired portions of the leases.

The action came on for trial before Judge Fay QC, the Official Referee, sitting as a deputy High Court judge, on the preliminary issue as to whether the action was statute-barred because more than six years had elapsed before the writ in respect of both the inspection and passing of the defective foundations, and the first sales of the flats.

At that time, *Sparham-Souter* had not been heard by the Court of Appeal and Judge Fay felt himself bound by the observations of Lord Denning in *Dutton* v. *Bognor Regis* (1972), the *fons et origo* of actions against local authorities for negligent supervision by building inspectors. He held that the plaintiffs' action was statute-barred.

However, after Lord Denning's observation in *Sparham-Souter* (see page 42–3), the plaintiffs received leave to appeal out of time. The Court of Appeal then reversed Judge Fay, held that the plaintiffs' action was not statute-barred and re-asserted the discoverability test. The Council received leave to appeal to the House of Lords on terms that the order for costs made in the High Court and Court of Appeal should not in any event be disturbed, no doubt in the hope of settling once and for all when a cause of action accrued in the case of defective buildings.

Before the House of Lords it was argued for the Council: firstly, that Lord Denning had been right in *Dutton* in holding that the cause of action arose when the negligent inspection took place; secondly, that if it were not then, the principle of *Cartledge* v. *Jopling* still applied and that the cause of action arose when damage was suffered by the plaintiffs, whether they knew of it or not, and that was when the defective foundations were constructed; or alternatively, when they bought apartments in a block with defective foundations.

Those contentions were all dismissed by Lord Wilberforce giving the leading opinion, endorsed by three other law lords. He began by asking 'when does the cause of action arise?'

He said: 'We can leave aside cases of personal injury or damage to other property, as presenting no difficulty.' In this fashion, he dismissed *Cartledge* v. *Jopling* and the subsequent statutes as of no relevance whatsoever to the issues before the court. He then continued:

> 'In my respectful opinion the Court of Appeal was right when in *Sparham-Souter* v. *Town and Country Developments (Essex) Ltd* (1976) it abjured the view that the cause of action arose immediately upon delivery i.e. conveyance of the defective house.'

Although that was reference to only one aspect of the Court of Appeal case, he made no criticism whatsoever of the judgment in *Sparham-Souter* and it is quite apparent that he accepted and endorsed it.

However, he reformulated the discoverability test by saying that the cause of action can only arise 'when the state of the building is such that there is a present or imminent danger to the health or safety of persons occupying it'. There can be no doubt that he assumed that, by the time the building was in that state, the occupiers would know it. In effect, he was putting the time even later. It was not when the first cracks appeared, which might put the occupiers on inquiry, but later when there was, by an objective test, 'a present or imminent danger to the health or safety of the persons occupying it'. In short, he seems to have dropped the concept that the cause of action arose when the plaintiffs could, by reasonable diligence, have discovered the defects; and the concept that they had to sue within six years of the discovery that there were defects 'which *at some future but uncertain date* could cause injury to health or safety'.

Lord Wilberforce emphasised that he was dealing only with the issue as to when a cause of action accrued. 'A cause of action arises at the point that I have indicated.' He then went on to point out that a plaintiff did not actually have to wait until the building was falling down over his head before issuing the writ, but if repairs could be carried out to prevent that happening, he was entitled then to do so.

The case returned to Judge Fay with the direction that the action was not statute-barred, since the first cracks appeared only in 1970 and the writ was taken out only in 1972, even though the flats had been first sold in 1962, the action continued for a time before him on other grounds, but was eventually settled on terms which were not disclosed in court and which the plaintiffs undertook not to disclose.

Inspection of the Council's minutes, however, revealed that the settlement involved the Council buying the whole of the block from the leaseholders on a 'market value as sound' basis, paying their entire costs on a solicitor and own client basis and offering them protected tenancies for life if they so desired. It had been an expensive operation for the Council, not all of which was met by their insurers.

The House of Lords' judgment in *Anns* caused a minor uproar in the construction industry, in particular amongst architects. The *RIBA Journal* reported that in the course of a year one in three of all practising architects in the country were likely to receive writs. Other local authorities were critical of Merton for taking the case to the House of Lords. They correctly concluded that it had made the situation worse than *Sparham-Souter*, not better, in that no longer could they raise the defence that the defects could have been discovered earlier if reasonable diligence had been exercised.

Damage for financial loss alone

Before the next round in the limitation saga, the House of Lords heard the appeal in *Junior Books* v. *Veitchi Co. Ltd* (1982), a Scottish action in delict (tort) against the nominated flooring sub-contractor of a building. The law of England and Scotland were said to be the same for the purpose of this action.

Two years after the floor had been laid, it developed cracks which did not constitute a danger to the health and safety of the occupiers but did involve the plaintiffs in a cost of some £220,000 in closing their factory down and relaying the floor.

The House of Lords held, *inter alia*, that:

(a) the sub-contractors owed a duty to the employers in tort to exercise due care, since there was 'a proximity' between them although no contractual relationship
(b) an action for a negligent act or omission does not require physical damage: financial loss alone is enough
(c) the duty of care in a tort or delict action extends beyond 'a duty to prevent harm being done by faulty work' to persons or other property, to 'a duty to avoid such faults being present in the works themselves'
(d) the 'pursuers' (plaintiffs) therefore had a cause of action against the sub-contractor in the delict (tort) of negligence for the financial loss alone.

The implications of this decision on the limitation periods were not discussed and in any event they were subject to Scottish law [2.10]. But in the course of his opinion, the eminent commercial lawyer, Lord Roskill, expressly approved of the *Anns* case.

The implications of this case are that a cause of action will arise in tort when it becomes necessary, for commercial reasons, to rectify defects discovered in work done by a contractor or sub-contractor. This is a quite separate and distinguishable cause of action. It might apply, for example, in the case of all those numerous houses in the North-east of England, the foundations of which have been infilled with 'shale' — not the oil bearing sands but refuse from steel furnaces. No physical damage has yet taken place, the National House-Building Council refuses to indemnify the owners unless it has occurred within the ten year period, but they have nevertheless suffered financial loss because their houses are on an estate which is 'blighted' and unsaleable.

There were various barbed comments in the opinions of the House of Lords in the *Junior Books* case as to why the main contractor had not been sued. The present author can disclose the reason: firstly, as a result of other rectified defects, the contractor had been given a total and absolute discharge; secondly, he was not worth suing and shortly afterwards went into liquidation.

The effect of the decision in the *Junior Books* case must surely be that a plaintiff can sue in tort when he is likely to be involved in financial loss as a result of the need to rectify defective workmanship, even though this does not involve 'present or imminent danger to health or safety'; but he is not *obliged* to do so but can wait for a separate and later cause of action when the works *do* involve this.

What did 'Pirelli' decide?

Later the same year, *Pirelli General Cable Works Ltd* v. *Oscar Faber & Partners* (1982) surfaced in the House of Lords. It had all the hallmarks of a contrived action; there had been a prior agreement as to costs between the parties; the action passed through the Court of Appeal by agreement between the parties like a dose of cascara. The plaintiffs conceded that more than six years before the writ had been issued there had been damage to the chimney in question. How they knew that was never explained.

The case concerned an industrial chimney erected under the defendants' supervision between June and July 1969 for the plaintiffs. It was designed and built by specialist sub-contractors who, for reasons not explained in the judgments, were not available to be sued. The action was therefore brought against the consulting engineers, Oscar Faber & Partners, who had approved the design. They denied liability for a sub-contractor's design and raised the issue of limitation. The plaintiffs conceded that cracks in the chimney occurred not later than April 1970 but they did not discover the defects until November 1977: the writ was issued in October 1978. It appeared to be common ground that the first date on which the plaintiffs could, with reasonable diligence, have discovered the cracks was October 1972 — just within the limitation period of six years.

Lord Fraser of Tullybelton gave the opinion of the House. He took the view that, in the *Anns* case, the House of Lords had not approved the 'discoverability test' of *Sparham-Souter* and that, following *Cartledge* v. *Jopling*, the cause of action arose when the damage occurred, whether the plaintiff knew of it or not. The 'discoverability test' of *Sparham-Souter* was expressly disapproved of, the 'danger to health or safety' of *Anns* was ignored: the first appearance of cracks, even if the plaintiff did not and could not possibly have known of them, was regarded as the date when the 'cause of action' accrued.

His Lordship went even further. He held, without adequate explanation, that the Limitation Act applied to building owners 'as a class'. So that the first owner, who did not know he had a defective building, had time running against him from the moment there was damage, and that all subsequent owners were from that time statute-barred. He also expressed the view that if the building was 'doomed from the start', time began to run from the moment that it was constructed. It followed that the more negligent a designer or contractor was, the earlier he was relieved from liability.

This was not a conclusion which commended itself to everybody for reasons which Lord Roskill set out in *Sparham-Souter*, but the decision in *Pirelli* was greeted with paeans of praise from architects, civil engineers and local authorities. Cambridge University abandoned its several million pound action on its new History Faculty building and the receiver for the Metropolitan Police gave up his action against the architects, the civil engineers and the contractors of New Scotland Yard which had major cladding defects.

Lord Fraser held that the test laid down by Lord Wilberforce of 'present or imminent danger to health or safety' was related exclusively to the duty resting upon local authorities, which was different from the duty resting on builders or architects. That was in spite of Lord Wilberforce's expressed views that it was equally applicable to builders.

Subsequently the Court of Appeal in *Ketteman and Ors* v. *Hansell Properties Ltd and Ors* (1984) disowned Lord Fraser's observations about plaintiffs 'as a class' and the 'doomed from the start' theory, but the English courts were left nevertheless with three inconsistent decisions by the House of Lords: *Anns, Junior Books* and *Pirelli.* Not surprisingly, they chose to follow the most recent, *Pirelli.* So the law was apparently left in the illogical position that:

(1) against the builder, the architect and the engineer time would begin to run when physical damage *first* occurred to the building, whether the plaintiff knew of it or not;
(2) in the case of financial loss caused by the necessity to repair

defects, the cause of action accrued when money had to be expended to repair the defects;

(3) in the case of actions against local authorities it accrued only at a much later time — when the defects constituted 'a present or imminent danger to health or safety'.

And in any event, these decisions begged the essential question: when is damage suffered?

When is damage suffered?

The present writer takes some pride in recalling that while practising at the Bar more than a quarter of a century ago, he recovered damages for injured workers from their employers by what became known on the North-eastern Circuit as the 'Parris Plea'.

There was in that area a factory where baths were made. Because of the conditions in the factory, many of the workforce suffered from lung disease and quite a number died prematurely in agony. If any sued, they were met with the limitation plea that the disease had been suffered more than six years before the issue of the writ, i.e. well before they knew that they had a cause of action.

The 'Parris Plea' was simple. It was an endorsement on the writ:

> 'The plaintiff claims damages for the injury, suffering and loss sustained by him in the six years immediately preceding the writ herein by reason of negligence and breach of statutory duty of the defendants, his employers, as more particularly set forth in the statement of claim delivered herewith.'

The statement of claim made it clear that the action did not seek to recover damages for injury, suffering or loss incurred before the limitation period. Needless to say, the injury, suffering and loss in the six years preceding the writ were very much worse than in the earlier period. There was one trial which, since it took place on circuit, was not reported, and no appeal to the Court of Appeal. Thereafter, as soon as they got the writ, the employers (or their insurers) paid up. Justice was done.

It may be of some interest that, much more recently, two New Zealand courts have decided that in the case of progressive damage to buildings, such a plea can be successful: the plaintiff can sue and recover in respect of the damage suffered within the preceding six years, even though there has been earlier damage: *Bowen and Ano.*

v. *Paramount Builders and Ano.* (1977); *Mount Albert Borough Council* v. *Johnson* (1979). It is well settled that, in actions for nuisance, this is the position, and the learned editor of *Salmond on Tort* says in respect of negligence:

> 'When an action is actionable only on proof of actual damage, successive actions will lie for each successive and distinct accrual of damage.'

Limitation periods in the Republic of Ireland

Meanwhile, Miss Justice Carroll in the Irish High Court rejected the *Pirelli* decision, in a masterly and convincing judgment which was not appealed against: *Brian Morgan* v. *Park Developments Ltd* (1983). The Irish statute in all material respects was the same as the Limitation Act 1939, which in turn is the same as the Limitation Act 1980. Normally, the Irish court, like most common law Commonwealth courts, applies decisions of the House of Lords.

The Law Reform Committee recommendations

In the meantime, the Law Reform Committee has published recommendations that in the case of latent defects the statutory limitation period should run from the date of damage or for three years from the time when the plaintiff discovers the defect, but that there should be a fifteen year 'cut-off' period from the date of the wrongful act or omission, after which no action should be possible.

As the law stands when the second edition of this book went to press, the limitation period in tort begins to run when damage is suffered by a building whether the owner or occupier knows of it or not. There is, however, great uncertainty as to when the damage is suffered. So far as local authorities are concerned it was held in *Billam and Ano.* v. *Cheltenham Borough Council* (1985) that it is not when the first cracks appeared as the result of defective foundations, but when the house became an imminent danger to the health or safety of the occupiers. This case is currently under appeal.

It is not a satisfactory position, and adoption of the Law Reform Committee's recommendations will make it no better.

Meanwhile the burden of proof rests on the defendant to establish that damage occurred outside the limitation period: *London Congregational Union* v. *Harriss* (1984); *Perry* v. *Tendring District Council* (1984).

2.14 The effect of the Civil Liability (Contribution) Act 1978

The relationship between the employer and the contractor, the contractor and sub-contractor and suppliers, and all these with the architect, are subject to the provisions of the Civil Liability (Contribution) Act 1978.

Formerly by section 6 of the Law Reform (Married Women and Tortfeasors) Act 1935 (now repealed), the court had power to apportion damages awarded to a plaintiff against joint tortfeasors:

> Section 6(1) 'Where damage is suffered by a person as a result of a tort . . . (c) any tortfeasor liable in respect of that damage may recover contribution from any other tortfeasor.'

The contribution recoverable was by:

> Section 6(2) '. . . such as may be found by the court to be just and equitable having regard to the extent of that person's responsibility for the damage.'

That Act did not authorise contribution between parties unless they were joint tortfeasors in respect of the same tort, and if one party was liable only in contract and the other in tort, there was no power to apportion damages. It raised enormously difficult questions of law, as for example where one tortfeasor was proceeded against and succeeded in establishing that the action was statute barred against him, as in the House of Lords' case of *George Wimpey Ltd* v. *BOAC* (1954). When it was applied to an employer's liability to his employees and it was discovered that the insurers of the employer could secure a contribution amounting to a complete indemnity from the employee who was negligent, it created a furore: *Lister* v. *Romford Ice and Cold Storage Co. Ltd* (1957). In practice, for many years this aspect of the statute was not utilised because of a 'gentleman's agreement' by employers' organisations and insurers not to rely upon it. But fears of this use were revived by a bank's successful action against one of its managers for contractual negligence in authorising an overdraft: *Janata Bank* v. *Ahmed* (1981).

Under the Law Reform Act, there could be no apportionment of liability in respect of a cause of action framed in contract (e.g.

against an architect or builder) or against a local authority for negligence in passing defective work.

The Civil Liability (Contribution) Act 1978 changed all that.

Under this Act 'any person liable in respect of any damage suffered by another person may recover contribution from any other person liable in respect of the same damage (whether jointly with him or otherwise)': section 1(1). A building owner can sue his architect, who can bring in for contribution the builder; or vice versa.

The assessment is to be, once again, such as the court finds 'just and equitable having regard to the extent of that person's liability for the damage in question': (section 2).

But the Act does more than bring in for contribution any cause of action.

Section 1(4) allows 'a person who has made any payment in *bona fide* settlement' to recover contribution 'without regard to whether or not he himself is or ever was liable in respect of the damage'. In other words, it is not necessary for a party seeking contribution to prove that he himself was in fact liable.

A new section 4 is substituted in the Limitation Act 1963 (time limit for claiming contribution) which limits the right to contribution to two years 'from the date on which that right accrued'.

All of it must be applicable, irrespective of the wording of the various contracts between architects, contractors, sub-contractors and suppliers. It is likely to increase the readiness of owners of defective buildings to sue everybody in sight, including inevitably the local authority, and to leave the court to apportion damages, irrespective of the terms of the contracts, on the basis of what the court considers 'just and equitable'.

It has also been held that even if one defendant settles with the plaintiff, another defendant is still entitled to claim a contribution from him if the appropriate notice has been served.

In *Logan* v. *Uttlesford District Council and Ors* (1984), the architect who had settled with the plaintiff in an action for damages was brought in as a third party for a contribution by another defendant. Application was made to a Queen's Bench master to strike out the third party notice and statement of claim and was refused. On appeal to a judge, Mr Justice Sheen upheld the decision, saying 'the settlement in the present case does not come within the proviso in sub-section 1(3) of the Civil Liability (Contribution) Act 1978'.

Already, as a result of the doctrine of dual liability in contract and tort in building cases, a range of parties may be sued. In *Acrecrest Ltd* v. *W. S. Hattrell & Partners and Ano.* (1979) a block of flats proved defective as the result of heave caused by the removal of trees. The architects were sued by the building owners, who brought in the local authority as joint tortfeasors by reason of the negligence of the building inspector. Sir Douglas Frank QC sitting as a deputy judge of the High Court assessed the contribution of the local authority at 25 per cent. In *Eames* v. *North Herts District Council* (1980) warehouses were built on an unconsolidated rubbish dump. Damages were apportioned between the developers, the builders and the foundation sub-contractor, all at 22.5 per cent of the damages. The architect was held liable for 32.5 per cent of the damages. This was an action framed in contract and tort but the apportionment was because all parties were held to be joint tortfeasors.

Although the limitation aspects of both these cases were disapproved of by Lord Fraser in the *Pirelli* case, this does not affect the principles of contribution.

The Act may not be all that important as the result of the dual liability in contract and tort. In *McGuirk and Ano.* v. *Homes (Basildon) Ltd and Ano.* (1982), the developer, who had a contract with the plaintiffs, and the builder, who had not, were found jointly liable to the plaintiff for using defective material of one part cement to thirteen parts of sand on a house that had been built twenty years before. Damages were assessed for the defective material at more than twice what this house had been built for originally. A defence that the claim was statute barred was pleaded but not pursued before the court.

But there have apparently been a flood of claims for contribution against architects by contractors and sub-contractors, who have settled claims made against them by employers, as the result of the Act.

2.15 Lump sum contracts

Contracts where a contractor undertakes to do work for a fixed sum are, as a matter of principle, subject to the rule in *Cutter* v. *Powell* (1795): nothing is due until the whole of the work has been completed. The second mate of a ship had contracted to sail from Jamaica to Liverpool for the sum of thirty guineas. He served for seven weeks on the voyage but just before the ship docked at Liverpool, he died. It was held that his widow was entitled to nothing since he had not performed expressly what he had undertaken.

In fact, the contract was a unilateral one whereby the defendant undertook: 'Ten days after the ship *Governor Parry* arrives at Liverpool I promise to pay Mr T. Cutter the sum of thirty guineas, provided he proceeds, continues and does his duty as second mate from here to the port of Liverpool.' But from it has been derived the concept of an 'entire' contract where one party's obligations have to be entirely fulfilled before he is entitled to any payment whatsoever.

Similarly, in the last decade of the last century, a builder undertook to build a house for £1565, and abandoned the project half way through. It was held that he was entitled to nothing for the work he had done: *Sumpter* v. *Hedges* (1898).

That harsh rule still applies in principle today: *Ibmac Ltd* v. *Marshall Ltd* (1968). But it has been mitigated so far as building work is concerned by the doctrine of 'divisible contracts' as in *Hoenig* v. *Isaacs* (1952) and 'substantial completion': *Dakin* v. *Lee* (1916).

Had the roles been reversed in *Sumpter's* case, so that it was the employer who had refused to let the builder complete, the builder had the option of charging for the work on a *quantum meruit* basis [2.16] or of claiming damages which would include all the profit he would have made on the job.

Builders frequently believe that even if they do not comply exactly with the contract, if they confer some, or a similar, or a greater benefit on their employer, they are entitled to be paid. They fall into the same error as did a shipyard which did repairs to a ship, under a lump sum contract, in a case called *The Liddesdale* (1900). They did not comply with the contract specifications, but did work which was more expensive and used materials which were more suitable. It was held that they could recover nothing. English contract law is based upon promise, not on benefit conferred, and the equitable restitution is applicable only where the defendant has an option to accept or reject. 'If a man, unsolicited, cleans my shoes, what can I do but put them on?' remarked a judge in an old case.

Of course, it is possible for any contract to be varied with the consent of the other party or for any departure from specification to be ratified subsequently; but in the absence of either, a contractor is entitled to nothing even though he may have expended considerable sums and enriched the owner.

2.16 Quantum meruit

The words '*quantum meruit*', which mean literally 'as much as is deserved', are often used instead of '*quantum valebant*', i.e. 'as much as something is worth,'.

The words are used in three different senses:

(1) Where there is a contract for the supply of goods or services, but the parties have not agreed how much shall be paid for them.
(2) Following a breach of contract by one party, when the other is prevented from earning a lump sum promised, he is entitled to the value of the work he has done.
(3) In equitable situations, apart from contracts, where compensation would be appropriate.

So far as contractual situations are concerned, Lord Dunedin pointed out in *The Olanda* (1917):

> 'When there are two parties who are under a contract, *quantum meruit* must be a new contract and in order to have a new contract, you must get rid of the old contract.'

In *Gilbert & Partners* v. *Knight* (1968) building surveyors undertook to supervise alterations to the defendant's house for the sum of £30. The alterations, originally estimated at £600, finally came to £2283. They then sent the defendant a bill for £135 — the £30 originally agreed and 100 guineas they reasonably thought they were entitled to for extra work. They recovered nothing. The Court of Appeal held there had been no fresh agreement and therefore there were no circumstances in which a promise to pay a *quantum meruit* could be implied.

There are many occasions when architects allow contractors to get away with similar claims to which they are not entitled. An architect cannot certify for anything other than what the JCT contract authorises him to do, and cannot therefore authorise *quantum meruit* payments, unless he has the express authority of the employer.

The JCT 80 contracts purport to be entire contracts for a lump sum. The actual effect will be discussed in connection with payment on final accounts [11.07].

2.17 A judicial summary of the law

In *Holland Hannen & Cubitt (Northern) Ltd* v. *Welsh Health Technical Services Organisation and Ors* (1981), Judge John Newey QC, sitting as an Official Referee, provided an admirable summary of the law relating to the JCT 63 contract, which is equally applicable to JCT 80.

(1) An entire contract is one in which what is described as 'complete performance' by one party is a condition precedent to the liability of the other party: *Cutter* v. *Powell* (1795) and *Munro* v. *Butt* (1858).
(2) Whether a contract is an entire one is a matter of construction; it depends upon what the parties agree. A lump sum contract is not necessarily an entire contract. A contract providing for interim payments, for example, as work proceeds but for retention money to be held until completion is usually entire as to the retention monies, but not necessarily the interim payments: Lord Justice Denning (as he then was) in *Hoenig* v. *Isaacs* (1952).
(3) The test of 'complete performance' for the purpose of an entire contract is in fact 'substantial performance': *H. Dakin & Co. Ltd* v. *Lee* (1916).
(4) What is 'substantial' is not to be determined on a comparison of cost of work done and work omitted or done badly: *Kiely & Sons Ltd* v. *Medcraft* (1965), and *Bolton* v. *Mahadeva* (1972).
(5) If a party abandons performance of the contract, he cannot recover payment for work which he has completed: *Sumpter* v. *Hedges* (1898).
(6) If a party has done something different from that which he contracted to perform then, however valuable his work, he cannot claim to have performed substantially: *Forman & Co. Proprietary* v. *The Ship 'Liddesdale'* (1898).
(7) If a party is prevented from performing his contract by default of the other party, he is excused from performance and may recover damages: *dicta* by Mr Justice Blackburn in *Appleby* v. *Myers* (1866); *Mackay* v. *Dick* (1880).
(8) Parties may agree that, in return for one party performing certain obligations, the other will pay to him a *quantum meruit*.

(9) A contract for payment of a *quantum meruit* may be made in the same way as any other type of contract, including conduct.

(10) A contract for a *quantum meruit* will not readily be inferred from the actions of a landowner in using something which has become physically attached to his land: *Munro* v. *Butt* (1858).

(11) There may be circumstances in which, even though a special contract has not been performed there may arise a new or substituted contract; it is a matter of evidence: *Whittaker* v. *Dunn* (1887).

Chapter 3

The role of the architect

3.01 Not a party to the JCT contract

The architect is not a party to the JCT 80 contract, although it repeatedly says he *shall* do this and *may* do that. There is therefore [2.03] no contractual obligation on him to do any of these things, except insofar as the contract may be annexed by reference to his Conditions of Engagement with the employer. Certainly, the contractor can never hold him *contractually* liable for failure to do anything that the JCT 80 says he shall do; but, of course, he may be liable to the contractor in tort.

It follows, also, that he cannot be joined as a party to any arbitration arising out of JCT 80 without his consent.

The duties laid on an architect under JCT 80 serve two purposes: they delimit the architect's authority in relation to the contractor, and they delimit the area in which the architect is acting as the authorised agent of the employer.

In *Stockport Metropolitan Borough Council* v. *William O'Reilly* (1978) the issues before an arbitrator were: the extent to which the works were varied; the determination of the contractor's employment and whether that was valid under clause 25 of JCT 63 or amounted to a repudiation of the contract; and whether the contractor had repudiated the contract. The arbitrator failed to distinguish between acts done by the architect which were within the scope of his authority under JCT 63 and other acts, and orders given to the contractor which were unauthorised by the contract, and for which the employer could not be held liable, in determining whether the contractor had repudiated his obligations or not.

Since, in concluding that the contractor was justified in withdrawing from the site and treating the contract as repudiated by the employer, the arbitrator had failed to distinguish between those acts of the architect for which the employer was responsible and those for which the architect alone was responsible, this amounted to an error of law on the face of the award and the award was set aside by the High Court.

The architect as agent for the employer

Usually when an architect is certifying under the contract, he is exercising his professional role independent of his employer and not as agent for the employer — *Sutcliffe* v. *Thackrah* (1974) — although it had been suggested earlier in that case that, in issuing interim certificates for payment, he is doing so as agent of the employer: *Sutcliffe* v. *Chippendale* (1972).

However, the facts may be such that he is to be regarded solely as the agent of the employer, as was the case with the City Architect (later termed the Director of the Environment) who was the architect named under a JCT 63 contract between Rees and Kirby Ltd and Swansea Corporation for the erection of a housing estate: *Rees and Kirby Ltd* v. *Swansea Corporation* (1983). The architect failed to certify on a claim for 'loss and expense' under clauses 11(6) and 24(1)(a), (c) and (d) (now JCT 80 clauses 13 and 26).

Mr Justice Kilner Brown said in this case:

> 'The more interesting point is whether the architect was agent for the defendants, making them vicariously liable. . . .
>
> As I understand the law an architect is usually and for the most part a specialist exercising his special skills independently of his employer. If he is in breach of his professional duties he may be sued personally.
>
> There may however be instances where the exercise of his professional duties is sufficiently linked to the conduct and attitude of the employer that he becomes the agent of the employers, so as to make them liable for his default.
>
> In the instant case, the employers, through the behaviour of the council and the advice and intervention of the Town Clerk, were to all intents and purposes dictating and controlling the architect's exercise of what should have been his purely professional duty.
>
> In my judgment, this was the clearest possible instance of responsibility for the breach attaching to the employers.'

Local authorities who refuse to allow their employed architects to certify until, for example, the district auditors have approved the payment of a claim may be making themselves liable vicariously — in the case under review it was interest charges of £206,629.89 which the judge held were recoverable under clauses 11 and 24, following the decision of the Court of Appeal in *F.G. Minter Ltd* v. *Welsh Health Technical Services Organisation* (1980). His decision so far as that was concerned is very much open to doubt, but his Lordship was undoubtedly right to hold that an architect could, on the facts, cease to exercise his independent role as a professional man and become no more than the tool or agent of the employer.

The employer will be estopped from denying that the architect has his actual authority to do those things specified in JCT 80. To take but one instance: variations. An architect has no implied authority to vary the work contracted, for under the RIBA Conditions of Engagement clause 1.31 it is expressly stated that he may not make material alterations etc. 'without the knowledge and consent of the client'. But a contractor does not have to worry whether or not the architect actually has the consent of his client for any instructions issued within the terms of JCT 80; the employer will be estopped so far as the contractor is concerned from denying that such variations were made with his consent.

Failure to do what the contract requires of an architect may serve as evidence of negligence in an action by the contractor in tort against the architect.

The architect undoubtedly owes a duty in tort to see that a contractor or a sub-contractor does not suffer loss by his negligence.

In *Lomas* v. *Hill* (1979) a sub-sub-contractor was brought in to do some minor demolition work, which involved knocking down a wall and shoring up while specially designed steelwork was placed in position. When he came to demolish the wall it was found that the wall was not one but two, and there was a heavy fireplace above, so that the whole building was in imminent danger of collapse. It was held that there was a duty of care on the architect and the structural engineer towards the sub-sub-contractor to ensure that he did not suffer loss through their negligence.

3.02 The architect's duties to the employer

In spite of their length, the RIBA Conditions of Engagement are incomplete in the sense that they make little or no reference to many duties which the courts have decided over the years constitute part of the architect's duties to his client, the employer.

These duties apply whether or not the person who undertakes this work is an 'architect' registered with the Architects' Registration Council as established by the Architects Registration Act 1931 (as subsequently re-enacted). In the United Kingdom, any person whether trained or not, is entitled to practise as an architect provided he does not use that word to describe his business. But the fact that a person is not on the register does not impose on him lesser duties than those which are imposed on architects. In *Oxborrow* v. *Godfrey Davis (Artscape) Ltd and Ors* (1980) a firm of estate agents prepared, for the sum of £30 10s 0d, plans for a bungalow drawn up by one of their employees who described himself as an 'architectural draftsman'. They were held just as liable as a registered architect would have been when the strip foundations designed proved inadequate for a clay site with trees in proximity to the bungalow. They were held responsible for 40 per cent of damages of £5 591, the builders 40 per cent, and the local authority, which had passed the foundations, 20 per cent.

It is outside the scope of this book to examine all of an architect's duties in detail but they include the duty to advise his employer correctly about the law, and about which contract to adopt. Although he is not required to have the expert knowledge of a lawyer, he is expected to have a command of the law applicable to planning, building regulations and bye-laws: *Townsend Ltd* v. *Cinema News, David A. Wilkie & Partners (Third Parties)* (1959); and the rights of adjoining occupiers and owners, including rights to light: *Armitage* v. *Palmer* (1960). As with other professional men, he will be expected to keep abreast of developments in the law, including recent legislation and court decisions, and will be liable to damages for negligence to his client if he does not do so: *B. L. Holdings Ltd* v. *Wood* (1979). In particular, he will be expected to have a detailed and expert knowledge of the JCT 80 contract if that is used.

3.03 Amending the contract

It is undoubtedly the duty of an architect to advise his employer which contract to use.

In *Hudson's Building and Engineering Contracts*, 10th Edition (1970), page 146, the learned editor wrote:

> 'The time must be rapidly approaching when architects or legal advisers recommending the use without modification of some of

the forms of the contract in general use at present in the United Kingdom (in particular in RIBA forms) must be in serious danger of an action for professional negligence.'

The same author, I. N. Duncan Wallace, in his *Building and Civil Engineering Standard Forms* (1969), page xiii, repeated many of the criticisms that appeared in the 9th Edition of *Hudson* and added: 'certainly it is my opinion that no adviser of any private employer should allow the forms to be used without substantial amendment.' That was a reference to JCT 63.

It is only fair to add that no architect, as far as is known, has been held to be negligent for recommending the use of JCT 63; nor is any likely to be since the courts, however critical they may be of the wording of it, are unlikely to condemn *in toto* a document which had the common consent of representatives of public employers, contractors and sub-contractors and architects. If the contract as a whole is adverse to the interests of the employer, other forms of contract, or substantial amendments to the Standard Form, are likely to arouse the suspicions of contractors and result in much higher tenders. The architect is entitled to weigh these factors in advising which forms of contract should be used.

At the same time, architects should consider the criticisms of I. N. Duncan Wallace, and judicial comments on the 1963 Standard Form.

There are now however substantial dangers in amending JCT 80.

Substantial amendments may turn it into being 'the employer's standard form' of contract for the purposes of section 3 of the Unfair Contracts Terms Act 1977 and also result in it being construed *contra proferentem* the employer in the event of any ambiguity [1.06].

3.04 Consequences of amending JCT 80 clause 25

There is a horrendous pitfall for the architect who wishes to amend clause 25 of the contract, the extensions of time clause.

This clause still contains in JCT 80 clause 25.4.7 (previously JCT 63 clause 23(g)) the much criticised extension of time for the contractor for 'delay on the part of nominated sub-contractors or nominated suppliers'.

It also includes as mandatory the two clauses, now clause 25.4.10.1 and .2 which were formerly 'starred items' and therefore indicated as optional in JCT 63 clause 23. These require a contractor to be given an extension of time if he is unable for reasons beyond his control, to secure the labour or materials he needs.

Most architects (and even more employers) will regard these as objectionable for obvious reasons and will wish to delete them.

However, all the fluctuation clauses provide that fluctuations are only to be frozen at completion date if 'the printed text of Clause 25 is unamended and forms part of the Conditions' [1.07].

This does not prevent clause 25 being amended by the deletion of 25.4.10.1 and .2 or in any other way the architect or employer sees fit to do so.

But if clause 25 is amended at all, it is important to delete JCT 80 clause 38.4.8.1 or clause 39.5.8.1 or clause 40.7.2.1 as well. Inadvertent amendment of clause 25 will otherwise result in the contractor being able to recover fluctuations for a period of overrun however long.

This may well serve to illustrate the possible perils of amending the Standard Forms. The only other amendments the present author would consider are the deletion entirely of JCT 80 clause 28 for reasons which will appear later [9.06], and the deletion of the whole of clauses 38.4.8 or 39.5.8 or 40.7.2 as applicable, and the deletion of clause 26.6 which preserves a contractor's other rights and remedies, and substitution of a clause confining a contractor's rights to those specified by the contract. There is no good reason, in this author's view, why a contractor, by virtue of that clause, should have rights conferred on him in excess of those available at common law and also preserve his common law rights in addition.

3.05 A further fluctuation pitfall for architects

JCT 80 clauses 38.4.8.2, 39.5.8.2 and 40.7.2.2 constitute another hidden pitfall for the architect. Unless he adjudicates upon every written application by the contractor for extension of time within the period specified in clause 25.3.1.4, the employer's right to freeze the contractor's fluctuations on the due date for completion is lost.

No longer can the architect cheerfully spike applications for extensions of time until he sees how things turn out. He is now required to deal with them 'not later than 12 weeks from the receipt of the notice'. If he does not, the employer may become liable for what may turn out to be enormous increases in labour and material costs and will no doubt look to the architect for compensation. A prudent architect will therefore ensure that for every request for an extension he receives, a prompt reply is sent and he sends a request for further and better particulars of the claim. Time only starts to run when he has secured 'reasonably *sufficient* particulars and estimate': JCT 80 clause 25.3.1.

3.06 The architect's certificates under JCT 80

The certificates to be issued by architects fall into three classes:

(1) where JCT 80 specifies the time in which they are to be issued
(2) where the contract is silent as to when they are to be issued but by implication of law they must be issued within a reasonable time
(3) where the contract is silent as to when they are to be issued but no implication arises that they must be issued at any particular time, subject to the proviso that they must be issued while it is still within the architect's power, i.e. before he issues a valid final certificate.

The situation is further complicated and confused by the provisions of JCT 80 clause 25.3.3:

25.3.3 'Not later than the expiry of 12 weeks from the date of Practical Completion the Architect shall in writing to the Contractor either

.3.1 fix a Completion Date later than that previously fixed if in his opinion the fixing of such later Completion Date is fair and reasonable having regard to any of the Relevant Events, whether upon reviewing a previous decision or otherwise and whether or not the Relevant Event has been specifically notified by the contractor under clause 25.2.1.1; or

.3.2 fix a Completion Date earlier than that previously fixed under clause 25 if in his opinion the fixing of such earlier Completion Date is fair and reasonable having regard to any instruction requiring as a Variation the omission of any work issued under clause 13.2 where such issue is after the last occasion on which the Architect made an extension of time: or

.3.3 confirm to the Contractor the Completion Date previously fixed.'

This requires the architect to review the position about extensions of time, whether or not the contractor has applied for them.

This clause replaces JCT 63 clause 22, about which there was unresolved controversy:

> 'If the Contractor fails to complete the Works by the Date of Completion stated in the Appendix to these Conditions or with an extended time fixed under clause 23 or clause 33(1)(c) of these Conditions and the Architect certifies in writing that in his opinion the same ought reasonably to have been completed, then the Contractor shall pay or allow to the Employer a sum calculated at the rate stated in the Appendix as Liquidated and Ascertained Damages for the period during which the Works shall so remain or have remained incomplete, and the Employer may deduct such sum from any monies due or to become due to the Contractor under this Contract.'

One view of this clause was that it conferred on the architect what one might reasonably term an equitable jurisdiction to avoid imposing liquidated damages at the specified rate on the contractor, and that the architect could take into account not only his own previous extensions and consider whether they were sufficient, but also events which were not specified in clause 23 as grounds for which he could grant an extension. That view was certainly erroneous.

The fluctuations clause JCT 80 38.4.8.2 (and the equivalents in clauses 39 and 40) refer only to the architect's obligation 'in respect of every written notification by the Contractor under clause 25'.

One lawyer has asked: 'Do these provisions, read together, mean that the position as to fluctuations is that the employer can deduct in a later interim certificate sums that he has previously paid in fluctuations, if before the two dates the completion date has been varied by the architect?'

It appears so. The architect deals promptly, that is, 'not later than 12 weeks from receipt of notice' with all the contractor's applications for extensions. So the employer's rights to freeze are preserved.

Later, on his review under JCT 80 clause 25.3.3.2, the architect fixes an earlier completion date. The contractor has been paid fluctuations on an interim certificate. It must follow that he is no longer entitled to these and must refund them.

Applications for extension of time

The architect should be alert to the fact that JCT 80 imposes time schedules which, if not complied with, may have serious consequences for the employer, including loss of his right to have the work completed on the contractual date, so that time for completion becomes 'at large' and loss of the right to liquidated damages follows.

Clause 25.2.1.1 lays an obligation on the contractor to give written notice 'whenever it becomes reasonably apparent that the progress of the Works is being or is likely to be delayed'. That is irrespective of whether he is making an application for an extension of time. In that notice he is called upon to specify what, if any, are 'relevant events' for the purpose of that clause.

On receipt of such a notice, the architect should carefully check that it complies with the contract in that it gives:

(1) the cause or causes of the delay anticipated
(2) which event, if any, is claimed to be a relevant event
(3) the expected effect of any relevant event
(4) an estimate of the extent to which this affects the completion date.

The architect is required *upon receipt* to grant such an extension of time 'as he *then* estimates to be fair and reasonable'. That is, the operative date is the date of receipt of the notice, not the date of his subsequent notification to the contractor.

There is then a period of twelve weeks, during which the architect has to notify this in writing to the contractor. If less than that period remains before the completion date, it has to be notified before the completion date.

What must be included on this notice is set out in clauses 25.3.1:

25.3.1 'If, in the opinion of the Architect, upon receipt of any notice, particulars and estimate under clauses 25.2.1.1 and 25.2.2,
.1.1 any of the events which are stated by the Contractor to be the cause of the delay is a Relevant Event and
.1.2 the completion of the Works is likely to be delayed thereby beyond the Completion Date

the Architect/Supervising Officer shall in writing to the Contractor give an extension of time by fixing such later date as the Completion Date as he then estimates to be fair and reasonable. The Architect/Supervising Officer shall, in fixing such new Completion Date, state:

.1.3 which of the Relevant Events he has taken into account and

.1.4 the extent, if any, to which he has had to regard to any instruction under clause 13.2 requiring as a Variation the omission of any work issued since the fixing of the previous Completion Date,

and shall, if reasonably practicable having regard to the sufficiency of the aforesaid notice, particulars and estimates, fix such new Completion Date not later than 12 weeks from receipt of the notice and of reasonably sufficient particulars and estimate, or where the period between receipt thereof and the Completion Date is less than 12 weeks, not later than the Completion Date.'

Confusingly, applications for extensions of time have to be resolved not more than twelve weeks after practical completion: clause 25.3.3:

25.3.3 'Not later than the expiry of 12 weeks from the date of Practical Completion the Architect shall in writing to the Contractor either

.3.1 fix a Completion Date later than that previously fixed if in his opinion the fixing of such later Completion Date is fair and reasonable having regard to any of the Relevant Events, whether upon reviewing a previous decision or otherwise and whether or not the Relevant Event has been specifically notified by the Contractor under clause 25.2.1.1; or

.3.2 fix a Completion Date earlier than that previously fixed under clause 25 if in his opinion the fixing of such earlier Completion Date is fair and reasonable having regard to any instruction requiring as a Variation the omission of any work issued under clause 13.2 where such issue is after the last occasion on which the Architect fixed a new Completion Date; or

> .3.3 confirm to the Contractor the Completion Date previously fixed.'

What is the effect if the architect fails to comply with this time schedule? There can be little doubt about it, as a number of cases have demonstrated: *Fernbrook Trading* v. *Taggart* (1979); *Miller* v. *London County Council* (1934). In *MacMahon Construction Pty Ltd* v. *Crestwood Estates* (1971), the New Zealand judge, Mr Justice Burt, said:

> 'If upon the proper construction of the power to extend, it should appear that the power must be exercised within a period of time either fixed or reasonable, then a purported exercise outside that time is ineffective and there then being no date from which the liquidated damages can run, the building owner loses the benefit of that provision.'

It must follow that failure by an architect to comply with the time schedule laid down by JCT 80 clause 25 will be a breach of the architect's professional obligation to his client. He will also be in breach of his common law duty to exercise due care in his relations with contractor and sub-contractor.

Where there is no specified time for the issue of a certificate, it may be inferred from the contract that the certificate will be issued within 'a reasonable time'.

In *Hick* v. *Raymond and Reid* (1893), Lord Watson said:

> 'When the language of a contract does not expressly, or by necessary implication, fix any time for the performance of a contractual obligation, the law implies that it shall be performed in a reasonable time. The rule is of general application.'

In *Liverpool City Council* v. *Irwin* (1977), a term was similarly implied 'to make the contract work'.

It follows that when JCT 80 is silent about the time in which a certificate or instruction must be issued, a term may be implied that it is within a reasonable time.

Clause 1(2) of JCT 63, now clause 2.3 of JCT 80, was said by Judge John Newey in *R. M. Douglas Construction Ltd* v. *CED Building Services* (1985) to be one such clause. He said:

> 'Condition 1(2) requires the architect, on being informed of a discrepancy between contract drawings and bills, to "issue instructions in regard thereto". The condition does not say within what period the instructions should be issued, but any substantial delay might well disrupt the whole contract. In those circumstances it seems certain that an obligation to give the instructions within a reasonable time would be implied.'

In some cases, however, there will be no implied term that the certificate will be issued 'within a reasonable' time. This was held to apply to clause 27(d)(ii) of JCT 63 certifying default by a nominated sub-contractor: *R. M. Douglas Construction Ltd* v. *CED Building Services* (1985). There is no cause of action by the contractor against the sub-contractor until this certificate is issued, and it would appear that so long as the architect has not issued a final certificate, he can provide one under this clause at any time.

This is so even though possession of such a certificate is a condition precedent to a contractor's action against a nominated sub-contractor for damages for delay or disruption: *Brightside* v. *Mitchell Construction* (1975).

One consequence of this is that the limitation period begins to run not when the breach of contract takes place but when the certificate is issued: *R. M. Douglas Construction Ltd* v. *CED Building Services* (1985).

3.07 Power to postpone work

The architect is empowered by clause 23.2 to issue 'instructions in regard to the postponement of any work to be executed under the provisions of this contract'.

The words, on the face of it, may allow the architect to veto the order in which the contractor proposes to do any work, but it is one to be used sparingly since the contractor may be entitled to an extension of time under clause 25.4.5.1. [7.06], and loss and expense under clause 26.2.5, if 'the regular progress of the works or any part therof' has been materially affected [10.07]; and it may even lead to determination of the contract by the contractor under 28.1.3.4 [9.06].

Instructions to postpone under this clause may be given, it has been said, in effect by implication. An architect's nomination of a sub-contractor on that sub-contractor's 'quotation . . . and tendering conditions', which terms involved delay in the contractor's original planned programme of the work, was held by Mr Justice White and the Court of Appeal in *Harrison* v. *Leeds* (1980) to amount to a postponement under JCT 63 clause 21(2). Lord Justice Megaw explained what he saw as the situation in these words:

> 'In the absence of some valid reason to the contrary, that instruction would necessarily operate as a postponement instruction . . . for it would mean that the contractor was instructed to agree with the sub-contractor that the contractor's intended concrete works should be postponed for some 11 months while the sub-contractor did the steelworks in accordance with [his] conditions.
>
> While the . . . order does not expressly use the word "postpone" or expressly give instruction to postpone, the order to make a contract with sub-contractors containing a condition which necessarily involves postponement is an instruction within the contemplation of the condition [amounting to an order to postpone].'

However, with respect, the logic behind that is rather dubious and the opinion of the experienced and learned arbitrator, David Gardam QC, that the order did not constitute a postponement order is in this writer's view greatly to be preferred. The contractor's intended, but not expressed, order in which he could carry out the works may have been the basis of his tender but that was in no sense part of the contractual obligation he undertook; still less was his programme, even if the architect had seen it, a contractual document. As the learned arbitrator found as a fact: 'The claimants prepared an outline programme . . . but this programme did not form part of their tender.'

If the Court of Appeal decision in this case be right, it seems that any alteration whatsoever to some optimistic programme, prepared in secret by a contractor before tendering, would amount to implied instructions by the architect to postpone. That seems improbable since JCT 80 (as did JCT 63) entitles the contractor to loss and expense and even power to determine the contract for postponement of the works. To place this interpretation on JCT 80 clause 23.2 is to ignore the words 'under the provisions of this contract'. It is submitted that they must mean that the clause only applies where there are contractual provisions regarding the order or date on which the work is to be executed. It must also require a formal order, and not be a mere incident or consequence of some other instruction.

The words in clause 23.2 read:

> 'The Architect . . . may issue instructions in regard to the postponement of any work to be executed under the provisions of this Contract.'

The architect has no power under this clause to postpone giving the contractor possession of the site [6.04]. The JCT have recognised this by making special provisions in the Intermediate Form for delay in giving possession of the site. An amendment to JCT 80 is currently expected.

3.08 Power to have defective work removed

JCT 80 clause 8.4 provides that the architect may issue instructions in regard to the removal from the site of any work, materials or goods which are not in accordance with the contract.

In *Holland Hannen & Cubitt (Northern) Ltd* v. *Welsh Health Technical Services Organisation and Ors* (1981) a sub-contractor had installed windows in a hospital which let in water and which the court held were defective both in design and workmanship. The architects issued notices purporting to be under clause 6(4) of the 1963 form (which had the same wording) condemning the window assemblies as not complying with the contract. Judge John Newey QC said:

> 'The first of these documents was in letter form and did not mention clause 6(4). The second and third were more formal and referred to the sub-clause. None of them, however, instructed Crittalls to remove the window assemblies from the site. The effect of doing so would have been to expose the interior of the hospital to wind and rain. . . . In my opinion, an architect's power under clause 6(4) is simply to instruct the removal of work or materials from the site on the ground that they are not in accordance with the contract. A notice which does not require removal of anything at all is not a valid notice under clause 6(4).'

This is a strictly literal interpretation of the wording, but if a notice is given which is expressed to be under clause 8.4 (or JCT 63 clause 6), condemning the windows, it surely implies that the contractor is to remove the defective items which are not in accordance with the contract. However, since this decision may be followed, it is highly important that architects phrase their requirements of a notice issued under JCT 80 clause 8.4 in strict accordance with its wording.

The judge added:

> 'I think that the three purported notices in this case were all invalid and of no effect under the contract. If I am wrong and the notices are to be construed as having required removal of the window assemblies, then they were instructions which were not obeyed. Since [the employers] did not invoke their sanction under clause 2(1) of bringing other contractors to remove the windows, their effect has long since been spent.'

It would therefore appear, according to this court, that if an architect issues a notice under clause 8.4 which is ignored, and no further action is taken beyond allowing the contractor to attempt remedial work, this acquiescence will in time deprive the notice of all effect. With respect, it must be doubtful if this is good law.

3.09 No power to direct contractor

The 1939 and earlier RIBA contract forms authorised the architect to give 'directions' to the contractor. In spite of those unqualified words, it is clear that at all times 'it is the function and right of the contractor to carry out his building operations as he thinks': *Clayton* v. *Woodman & Son Ltd* (1962). The architect has no power to tell a contractor how to do the work, or in what sequence: *GLC* v. *Cleveland Bridge* (1984).

Moreover, a term is to be implied into the contract that the employer, and perforce his architect, will not interfere with the contractor's due progress.

In *Holland Hannen & Cubitt (Northern) Ltd* v. *Welsh Health Technical Services Organisation and Ors* (1981) it was pleaded on behalf of a contractor under a JCT 63 contract (July 1969 edition) that there were implied terms whereby the employer contracted that he and his architect:

> 'would do all things necessary on their part to enable [the contractor] to carry out and complete the works expeditiously, economically and in accordance with the contract'

and separately, that neither:

> 'would in any way hinder or prevent [the contractor] from carrying out and completing the works expeditiously, economically and in accordance with the main contract.'

Judge John Newey QC held that the failure of the architects to issue a variation order when they discovered that windows designed and installed by a nominated sub-contractor were of defective design was the employer's breach of these implied terms and also a breach of the architects' duty to their clients in contract and tort.

3.10 New power to vary

It will be noted, however, that clause 13.1.2 extends the meaning of the word 'variation' and makes a fundamental change from the provisions of JCT 63.

The architect now has power, if obligations are contained in the bills of quantities — but only if — to issue a variation order in respect of access to the site or any specific part of the site, limitations of working space or hours. It also gives him power in similar circumstances to vary the execution or completion of the work in any specific order. It must be stressed that this power exists only if the original bills of quantities contain provisions governing these matters and it would appear not applicable if reference were made to these matters in a specification or other documents extraneous to the bills.

The contractor has power to object to variations under this clause. That right is conferred by clause 4.1.1 which exonerates him from complying with a variation 'within the meaning of Clause 13.1.2'. He need not comply with the architect's instructions 'to the extent that he makes reasonable objections in writing to the Architect to such compliance'. If a dispute arises as to whether or not his objection is reasonable, the arbitration provisions in Article 5 allow immediate arbitration without waiting until after practical completion.

Is it a reasonable objection that, as a result, he will not make as much money as he would otherwise have made?

3.11 Correcting discrepancies

Clause 2.2 provides that discrepancies between the bills and SMM6, or in descriptions, quantities or omitted items shall be corrected and such corrections treated as variations. Although the architect is not specifically mentioned, it is he who has to do the correction.

Clause 2.3 deals with other kinds of discrepancies between the contract drawings, bills and architect's instructions.

Clearly, none of these provisions has any reference to a master programme prepared by the contractor and delivered to the architect in accordance with JCT 80 clause 5.3.12.

Clause 2.3.4 provides that on notification by the contractor the architect 'shall issue instructions in regard thereto'.

The previous provision in this respect was contained in JCT 63 clause 12(1) and (2). As with that clause, there is no provision whatsoever for the corrections of errors in pricing, or in multiplication or addition. If a contractor makes a unilateral mistake in his tendering he cannot claim to have the contract set aside on the ground that there is no *consensus ad idem*: *Higgin* v. *Northampton Corporation* (1927). In *Ewing and Lawson* v. *Hanbury & Co* (1900) an offer to work on metal at 30 shillings per cwt was carelessly accepted under the impression that it was 30 shillings per ton; it was held there was a contract at 30 shillings a cwt.

If, however, the contractor submits a tender containing a mistake against his interests and the employer or his architect or quantity surveyor sees it, deliberately keeps quiet and takes advantage of it, a court may well correct the error, apparently in exercise of its equitable jurisdiction and on the basis of estoppel: *Riverlate Properties Ltd* v. *Paul* (1975). The equitable doctrine of rectification of documents applies only when there is an error in the written expression of a previously concluded, oral or informal, agreement; or where there is an intention common to both parties to make a contract in terms different from those expressed on the document: *Joscelyne* v. *Nissen* (1970). However, in Canada where tenders are subject to bid bonds, the courts are apparently prepared to allow a contractor to withdraw his tender if, after the tenders are opened but before a contract is sealed, he discovers a major underpricing in his tender.

This is not, although it seems to profess to be, the position in English common law or equity.

3.12 Architect's instructions

Clause 4.3.1 requires that all instructions issued by the architect shall be in writing. It then goes on to give elaborate provisions to cover the situation where the architect gives only oral instructions. Clause 4.3.2 allows the contractor to confirm them within seven days and if within a further seven days from receipt of that notice the architect has not dissented from them, such instructions are effective. But from when? Not from the date of the original oral instruction but from the date which is seven days from that on which the architect received the contractor's notice.

Similarly, where within seven days of issuing an oral instruction the architect issues a written confirmation, that instruction is valid and effective only from the date of the confirmation: JCT 80 clause 4.3.2.1. In short, the original oral instruction is not retrospectively validated. If the contractor has complied with the oral instruction, it must be at his own cost, it would appear. Similar provisions apparently apply to instructions issued by a clerk of works: JCT 80 clause 17.

Strangely, if neither the contractor nor the architect confirm oral instructions but the contractor acts on them, the architect may (but is not obliged to) confirm them in writing at any time up to the issue of the final certificate.

In that case the instruction is validated not when he issues the confirmation but on the date when he originally issued the original oral instruction: clause 4.3.2.2.

3.13 The architect's obligations under JCT 80

The architect's duties, indicated by the use of the word 'shall' are:

Clause 2.3.4: Discrepancies
To issue instructions where the contractor has given notice specifying a discrepancy and divergence between the contract drawings, the bills of quantities or subsequent architect's instructions or drawings.

Clause 4.2: Specifying authority
To specify to the contractor in writing under which provision of the conditions he has issued an instruction if called upon to do so by the contractor.

Clause 4.3.1. Issuing instructions
To issue all instructions in writing.

Clause 5.2 and 5.3: Document copies
Once the contract has been signed or sealed, immediately to provide the contractor with one *certified* copy of the contract documents, two further copies of the contract drawings and unpriced bills of quantities and, as soon as possible, two copies of any descriptive schedules or 'other like documents' — unless the contractor already has these.

Clause 5.4: Further drawings
When necessary, subsequently to provide the contractor with two copies of further drawings.

Clause 5.8: Duplicate certificates
In respect of certificates issued to the employer, to send immediately a duplicate copy to the contractor.

Clause 6.1: Statutory requirements
To issue within seven days instructions regarding any divergence between contract documents and statutory obligations.

Clause 7: Determining levels
To determine any levels required for the work.

Clause 13.3: Provisional sums
To issue instructions regarding expenditure of provisional sums.

Clause 17.2: Defects liability
To specify and deliver a schedule of defects not later than fourteen days after expiration of defects liability period.

Clause 34: Antiquities
To issue instructions regarding fossils, antiquities and other objects of interest or value found by the contractor and to ascertain, or instruct the quantity surveyor to ascertain, any direct loss and/or expense occasioned thereby, unless the contractor is otherwise remunerated.

Chapter 4

The contractor's obligations

4.01 Implied terms

Before examining in detail the expression of contractual obligations contained in JCT 80, it is necessary to consider whether there are terms which the law will write into the contract. Such terms may be implied in this contract:

(a) by operation of law, or
(b) to give commercial effectiveness to the contract under *The Moorcock* doctrine [2.06].

If the implied term comes under the first category, it will apparently override contractual obligations to the contrary. In the House of Lords' case of *Young and Marten* v. *McManus Childs* (1969), the contractor complied exactly with the contract by using the specified Somerset 13 tiles, yet was held liable for latent defects in them [4.03]. In *Street and Ano.* v. *Sibbabridge and Ano.* (1980), an implied term that the contractor would comply with the Building Regulations was held to be an effective term of the contract, overriding the obligations to comply with the architect's design specification for the foundations.

In JCT 80 clause 6, of course, there are express terms regarding statutory obligations, such as the Building Regulations [4.08].

4.02 Implied terms: dwellings

The implied terms which were written into the Sale of Goods Act 1893 were those which existed in common law before the law relating to the sale of goods was codified into a statute. But there were, until recent years, no similar terms to be implied in building work. In part, this was due to the fact that the common law did not recognise buildings on land as anything separate from the land on which they stood.

So far as dwellings are concerned, and this includes blocks of flats as well as houses, the construction of these is now subject to the provisions of the Defective Premises Act 1972. This statute therefore imposes greater obligations on developers, contractors, sub-contractors, architects or engineers than are contained in JCT 80 and its accompanying documents and contracts. It has been suggested that even a local authority building inspector may be caught by the Act, as well as being liable in tort.

Section 1(1) enacts:

> 'Any person taking on work for or in connection with the provision of a dwelling . . . owes a duty to see that the work that he takes on is done in a workmanlike manner, with proper materials . . . and so as to be fit for the purpose required . . .'

Houses and flats built under the National House-Building Council scheme were once exempted from the provisions of this statute, by virtue of a statutory instrument, but have not been since 1975.

However, under HB5 there are express warranties which are independent one of another. In *Batty and Ano.* v. *Metropolitan Properties and Ors* (1978) a house was perfectly well built but on plateau land which, it was accepted by the trial judge, would slide down the hillside within ten years. It was contended that the express term 'so as to be fit as a dwelling house' was no more than the sum total of good workmanship and proper materials. It was held, however, that being fit for the purpose required was a requirement independent of, and overriding, these obligations.

For dwellings being built subject to the Act and under JCT 80, there is a curious and ambiguous defence contained in section 1(2). This reads:

> 'A person who takes on any such work for another on terms that he is to do it in accordance with instructions given by or on behalf of that other shall, to the extent to which he does it properly in

accordance with those instructions, be treated for the purposes of this Section as discharging the duty imposed on him by Sub-section (1) above except where he owes a duty to that other to warn him of any defects in the instructions and failed to discharge that duty.'

The Act specifies:

> 'a person shall not be treated for the purposes of Sub-section (2) above as having given instructions for the doing of work merely because he has agreed to the work being done in a specified manner, with specified materials or to a specified design.'

This curious situation leaves it open to argument that a contractor working under JCT 80 has no liability under the Act since he is required by clauses 2 and 4, amongst many others, to comply with 'all instructions issued to him by the architect' in regard to matters which the architect is by the contract authorised to issue.

It seems likely therefore that a builder who supplies and sells a finished product, where he is responsible for the design, materials and construction, to the 'instructions' of a lay client will not be able to rely on this defence. But a contractor working under JCT 80 with the employer's architect and with specifications, drawings and contract documents will be able to — provided of course, that he has complied with the architect's instructions, has given appropriate warning of any defects in those instructions, and provided a reasonably competent contractor would have perceived such defects.

4.03 Implied terms in construction contracts

Apart from statutory intervention, the courts have progressively implied in all building contracts terms that 'the builder will do his work in a good and workmanlike manner; that he will supply good and proper materials; and that it will be reasonably fit for the purpose required'. In *Hancock* v. *B. W. Brazier (Anerley) Ltd* (1966) a builder had used an infill which was commonly accepted as suitable but which, through no fault of his, reacted with the chemicals in the ground so as to cause heave. He was held liable for breach of the implied warranty. In *Young and Marten Ltd v. McManus Childs Ltd*

(1968) a roofing sub-contractor complied exactly with the employer's instructions to install 'Somerset 13' tiles, manufactured only by one maker, on the roof. He was held liable, when the tiles failed through a latent and undiscoverable defect, for breach of an implied warranty to supply good and proper materials.

In the case of *Test Valley Borough Council* v. *Greater London Council* (1979) the claimants' local authority predecessors entered into an agreement with the LCC, the respondents' predecessors, under powers conferred by the Town Development Act 1952, whereby the latter would erect and, when completed, sell to the claimants' predecessors some 6000 houses in Andover. The houses were intended to relieve over-population in London. The houses were duly completed and handed over, but many substantial defects subsequently appeared.

The matter came before the court on a case stated by an arbitrator dealing with claims made in respect of forty-four houses.

The substantial issue at the hearing was whether it was an implied term of the agreement that the respondents' predecessors would merely exercise reasonable skill and care in erecting the houses, or whether it was an implied term that the houses would be fit for human habitation. In other words: 'What standard of duty ought to be implied?' It was held by the High Court that:

> 'there were implied terms of the agreement that the respondents would not merely exercise reasonable care but would provide completed dwellings which were constructed (a) in a good and workmanlike manner; (b) of materials which were of good quality and reasonably fit for their purpose and (c) so as to be fit for human habitation.'

The question of implied terms in construction work again reached the House of Lords in the *IBA* v. *EMI and BICC* (1980), when the principles set out above were again approved.

Lord Fraser put forward two propositions in the course of that case:

> 'It is now well recognised that in a building contract for work and materials, a term is normally implied that the main contractor will accept responsibility to his employer for material provided by nominated sub-contractors . . . and the principle that applied in *Young and Marten* in respect of materials ought in my opinion to

> be applied here in respect of the complete structure, including its design.'

The structure was the 1250 ft cylindrical television mast on Emsley Moor, Yorkshire which crashed down on 19 March 1969.

The contract in that case was not the JCT form but the 'Model Conditions' of the Institution of Mechanical Engineers, the Institute of Electrical Engineers and the Association of Consulting Engineers, which contain limited obligations on the contractor not dissimilar to those in the JCT contract.

The defective television mast was designed and erected by BICC, nominated sub-contractor to EMI. EMI were held liable for breach of the implied terms as to fitness for purpose, written into their contracts by the courts, by the trial judge, in the Court of Appeal and the House of Lords.

It is noteworthy that the Model Conditions purport to exclude any warranties other than those expressly given by the contractor. It is not clear from the speeches in the House of Lords whether the attention of that court was specifically directed to this provision but, on the assumption that this was so, it would appear that the implied conditions imposed by law override any contractual provisions. This is the conclusion that Judge Fay arrived at in another case [4.09].

4.04 Implied terms in design and build contracts and sub-contracts

In the JCT contract 'with Contractor's Design' (1981) a deliberate effort has been made to reduce the contractor's liability in relation to design to that of professional men. Clause 2.5.1 of that contract reads:

> 'The contractor shall have in respect of any defect or insufficiency in such design the like liability to the employer, whether under statute or otherwise, as would an architect or, as the case may be, other appropriate professional designer holding himself out as competent to take on work for such design who, acting independently under a separate contract with the employer, had supplied such design for or in connection with, works to be carried out and completed by a building contractor not being the supplier of the design.'

In plain words, the contractor so far as his design element is concerned shall have no greater liability than that which is imposed by the *Bolam* standard of professional care. He does not warrant by this contract that his design will be reasonably fit for the purpose required.

A similar attempt is made in the nominated sub-contractor's standard contract with the employer, NSC/2, likewise to reduce the liability of the sub-contractor.

NSC/2 clause 1.1 provides that the sub-contractor:

> 'has exercised and will exercise all reasonable skill and care . . . in the design . . . selection of materials and goods . . . the satisfaction of any performance specification . . .'

That is, again, an attempt to exclude by implication the implied warranties as to good work and materials and fitness for purpose.

The question is: will the courts allow contractors and sub-contractors to get away with this reduction in their standards? It is most unlikely.

4.05 Implied terms in JCT 80?

The House of Lords considered this question in connection with the JCT 63 contract, where the obligations of the contractor were in similar terms to those in JCT 80, in *Gloucester County Council* v. *Richardson* (1968) shortly after their decision in *Young and Marten*. The case was concerned with whether a main contractor under JCT 63 was liable under an implied term for the fitness of materials supplied by a nominated supplier who had excluded the implied warranties of the Sale of Goods Act 1893. The five Lords of Appeal were then unable to agree on whether these implied terms were to be read into JCT 63. Two thought they were; two thought they were not; and one did not express an opinion. As the learned editor of *Hudson's Building Law* remarked: 'That case poses more questions than it answers.'

But those were early days in the development of the doctrine of implied terms in building contracts. There can be no doubt today that by operation of law any contractor under JCT 80 or the NSC/2 or NSC/4 contracts impliedly warrants that all that is done will be in a workmanlike manner, with materials of good quality (even if the materials are specified by the architect) and that these and the structure or any part of it will be reasonably fit for the purpose required.

These are overriding and strict obligations, and independent of fault. The contract terms may have been complied with exactly, the greatest care may have been exercised by the contractor, but he will be liable if any of these three warranties is breached.

The contractor's obligations under the JCT 80 contract are set out in clause 2.1:

> '[to] . . . carry out and complete the Works shown upon the Contract Drawings and described by or referred to in the Contract Bills . . . in compliance therewith, using materials and workmanship of the quality and standards therein specified, provided that where and to the extent that approval of the quality of the materials and of the standards of workmanship is a matter for the opinion of the Architect, such quality and standards shall be to the reasonable satisfaction of the Architect.'

This obligation carefully avoids any promise that the work will be done in a good and workmanlike manner, or that the materials will be of good quality and the complete work will be reasonably fit for the purposes required. Provided the work complies with the contract documents, the express obligation has been complied with. To compare with the case of *Young and Marten Ltd* v. *McManus Childs* already mentioned, if the contract documents require Somerset 13 tiles and these are supplied, the contract has been fulfilled.

In the present author's view, it is inevitable that the courts will hold in future that all three terms are to be implied into JCT 80.

In *Holland Hannen & Cubitt (Northern) Ltd* v. *Welsh Health Technical Services Organisation and Ors* (1981), it was pleaded by the defendants in a case on the JCT 63 local authorities contract (July 1969 version) that there were implied terms of the contract that the building work done by the contractor and sub-contractors under it should be carried out 'in a good and workmanlike manner and with proper care and skill', and that 'the materials used . . . should be of good quality', and that 'the materials used . . . should be reasonably fit for the purpose for which they were required'.

These terms were not disputed on the pleadings by any of the parties and the judge accepted them as being part of JCT 63 and the NFBTE/FASS sub-contractors' Green Forms.

It would appear therefore that the position now is that common law will imply these standard terms into all contracts irrespective of their terms; even if a contract were to expressly exclude them, which

JCT 80 does not, the exclusion clause would be caught by section 7 of the Unfair Contract Terms Act 1977.

The Law Commission in its 1979 Report of 'Implied Terms in Contracts for the Supplier of Goods' (Law Com. No.95) said in paragraph 66:

> 'Our conclusion . . . is that the supplier's obligations in respect of the materials supplied should be the same whether he simply sells them or whether he supplies them and also does work under the contract. In other words, the person who supplies goods under a contract for work and materials should be bound by the same implied terms in respect of the materials supplied in accordance with the Sale of Goods Act Model.'

The Supply of Goods and Services Act 1982 now contains statutory provisions on this point, but in effect the judges had legislated already.

4.06 Contractors' obligations under JCT 80 clauses 2 and 8

In addition to the obligations which the law will impart into the contract, clause 2 contains four separate obligations on the contractor. They are:

— 'to complete and carry out the works . . . in accordance with the contract documents'
— 'using materials . . . of the quality and standards therein specified'
— 'using . . . workmanship of the quality and standards therein specified'
— 'where the contract documents expressly make either quality or standard of these subject to the architect's approval, then there is an obligation that they will be to the "reasonable satisfaction of the architect".'

The last obligation deserves some amplification. Where there is such an exceptional term in the contract documents, the contractor must comply with it. The architect does not have to exercise his powers under clause 8 to require the removal of any materials or workmanship which does not meet with his approval. It is sufficient if he indicates that it is not to his satisfaction. Likewise, he is under no obligation to issue a variation notice to secure replacement of materials of which he disapproves. The architect is under no obligation to include in interim certificates work which is not to his satisfaction.

The contractor can have no claim of any kind for expense or loss caused by the removal and replacement of materials or workmanship which have not been to the architect's satisfaction, for it is the contractor who is in breach.

JCT 80 clause 8 appears in part to be unnecessary and repetitious, since it provides that:

> 'All materials, goods and workmanship shall be . . . of the respective kinds and standards described in the Contract Bills.'

Clause 2 has already said that. However, it introduces an inconsistency which may be of some importance. Materials, goods and workmanship have only to be provided 'so far as procurable'.

Clause 2 imposes an obligation on the contractor to provide the materials and workmanship specified in the contract bills. If he does not he is in breach of his contract. It is no excuse that they may be difficult to obtain or not readily available when he wants them. He has contracted to provide them.

By clause 8.2 he is under an obligation, if requested, to provide vouchers to prove that the materials and goods comply.

But clause 8.1 imposes quite a different obligation to clause 2. The contractor has only to provide them 'so far as procurable'. It appears to provide him with a ready-made excuse for not providing them, but does not go on to spell out what the effect should be if the materials, goods or workmanship should not be procurable. So far as goods and materials are concerned, it may at first glance appear to entitle him to treat substitution of other materials or goods as a variation under clause 13.1.1.2 provided they constitute 'alteration of the kind or standard of any of the materials or goods to be used'. This is definitely not what the industry understands. Substitution would not normally mean an alteration in the 'kind or standard'. The word 'kind' must mean of a different nature entirely; it is inappropriate for substitution of materials of a similar kind.

The only practical way is to treat such a substitution as a variation by agreement.

4.07 Contractor's master programme and other documents

It has been widely reported in the construction press that JCT 80 requires a contractor to produce a 'master programme' and to provide the architect with a copy of it. Even the JCT itself appears to believe that is what the contract says, for there is a footnote to clause 5.3.1.2 requiring that clause 'to be deleted if no master programme is required' and in the JCT's notes on major changes, reference is made to 'a new obligation on the provision of a master programme'.

That is not what the contract says. In fact, that clause does not impose an obligation on the main contractor to have a master programme. The structure of clauses 5.3.1.1 and 5.3.1.2 is exactly the same. The first requires the architect to provide the contractor with 'any descriptive schedules' and the second requires the contractor to provide the architect with a copy of his master programme — both clauses are effective only if descriptive schedules or a master programme exist. An architect would be unwise to insist on the contractor producing one and not only because he has no power to do so. However, it may be unlikely in modern conditions that the contractor has no master programme. He is not contractually obliged to have one, nor does the contract specify what it should comprise.

It has been suggested that, if the architect does require a master programme from the contractor, he should include such a requirement in bills of specification. But he is well advised not to specifically approve the master programme.

There is a change in JCT 80 about custody of the original contract drawings and contract bill. They are now to be retained in the custody of the employer under the local authorities' edition, but in the private edition remain with the architect: clause 5.1.

The contractor is under an obligation to keep on site a copy of the contract drawings, a copy of the unpriced bills of quantities, a copy of descriptive schedules 'or other like documents', a copy of his master programme (if he has one) and a copy of further drawings: clause 5.5.

The architect has power to require the contractor to return all documents which bear his name to him after final payment: clause 5.6. And the contractor in any event gives an undertaking that he will not use any of the documents for any other purpose than in connection with the contract in question: clause 5.7. It is difficult to see what use these provisions are or how an architect could enforce them, since he is not a party to this contract [2.02]; but no doubt the employer, if so minded, would be entitled to enforce them by obtaining an order for a mandatory injunction or for specific performance: *Beswick* v. *Beswick* (1968).

Notices the contractor is obliged to give

The contract provides for the contractor giving to the architect or employer various notices, some of which are mandatory. Others are merely directional, discretionary or conditional.

The mandatory notices to an architect appear to be:

Clause 2.3: *immediately* and in writing, upon discovery of any discrepancies between contract drawings, bills and instruction.

Clause 5.3.1.2: if he has a master programme, two copies to architect and of every subsequent amendment.

Clause 6.1.2: *immediately* and in writing, any divergence he has discovered between statutory requirements and contract drawings etc. or AIs.

Clause 6.1.4.2: *forthwith* of emergency compliance with statutory requirements.

Clause 22B.2: *forthwith* in writing (also to employer), notice of damage done by clause 22 perils.

Clause 26.4.1: upon receipt of written application by nominated subcontractor for loss and expense under clause 13.1 of NSC or NSC/4a.

Clause 30.6.1.1: before or within reasonable time of practical completion 'all documents necessary for purpose of the adjustment of the Contract Sum' including accounts of nominated sub-contractor and suppliers.

Clause 35.8: in writing, if unable within 10 working days to reach agreement with proposed nominated sub-contractor.

Clause 35.9: *immediately* in writing, if proposed sub-contractor named in tender NSC/1 withdraws.

Clause 35.13.3: before issue of each interim certificate (other than first) a reasonable proof of discharge of payments to nominated subcontractors.

The mandatory notices to an employer are:

Clause 19A.6 (local authorities edition only): to employer, *if required*, names and addresses of all sub-contractors.

Clause 21.1.2: as and when *reasonably required* documentary evidence that insurances required by clause 21.1 for personal injury, death etc. have been effected.

Clause 21.1.2: as and when *reasonably required* documentary evidence that insurances required by clause 21.1 for personal injury, death etc. have been effected.

Clause 21.2.2: deposit of insurance policies and receipt for premium of insurance under 21.2.1 for injury to 'any property other than the Works'.

Clause 22A.2: deposit of policies and receipts in respect of clause 22 perils or alternative in clause 22A.3.1.

Clause 22B.2: *forthwith* in writing (also to architect), notice of damage done by clause 22 perils.

Clause 31.3.1: 'not later than 21 days before first payment notice', notice of tax statutes under Finance (No. 2) Act 1975 and of any change in status *immediately*: clauses 31.4.1, 31.4.3.

4.08 Statutory obligations

JCT 80 clause 6.1 imposes on the contractor an obligation to comply with all statutory obligations. It also imposes an obligation on him by clause 6.1.2 to give immediate notice to the architect in writing if he discovers any divergence between the contract documents (including the drawings) and statutory obligations, including, of course, the Building Regulations.

This would also appear to cover among other things the situation where the contractor finds out that, because of the soil condition, the designed foundations will not comply with the Building Regulations.

That part of clause 6 ends with a proviso in 6.1.5:

> 'Provided that the Contractor complies with clause 6.1.2, the Contractor shall not be liable to the Employer under this Contract if the Works do not comply with the Statutory Requirements where

and to the extent that such non-compliance with the Works results from the Contractor having carried out work in accordance with the documents referred to in clause 2.3 or with any instructions requiring a Variation issued by the Architect in accordance with clause 13.2.'

In short, the clause purports to exonerate the contractor from any liability to the employer provided he carries out the architect's design.

It is submitted that this clause is ineffective to achieve its intended purpose. To start with, the clause only purports to exempt the contractor from his obligations 'under this contract'. Such words are insufficient to exclude the contractor from liability in tort for negligence in failing to exercise the reasonable skill and care of the ordinary, competent contractor. He will be liable to his employer if he carries out the architect's design when a competent contractor would have known that the design was defective, as in a Canadian case where the architect's design failed to provide for air vents in a timber house and the contractor was held liable for negligence in executing the architect's design since any competent contractor would have known it was defective.

Moreover, so far as the Building Regulations are concerned the obligation is on the contractor to comply with them absolutely and his employer may have a cause of action for breach of statutory duty [2.09] if he does not do so. This tortious liability is not excluded by JCT 80 clause 6.1.5.

Even if this clause by its wording excludes liability for breach of the express warranty given in clause 6.1, it may not be wide enough to exclude the implied warranty to comply with the Building Regulations.

In any event, clause 6.1.5 is clearly caught by section 7 of the Unfair Contract Terms Act 1977.

This applies to situations 'where the possession or ownership of goods passes under or in pursuance of a contract not governed by the law of sale of goods or hire purchase . . .'. That is, it applies to construction contracts. It applies to 'contract terms excluding or restricting liability for breach of obligation *arising by implication of law* from the nature of the contract'.

Section 7(2) then provides that 'as against a person dealing as a consumer, liability in respect of goods . . . quality or fitness for any particular purpose cannot be excluded or restricted by reference to any such term', and only in other circumstances, if reasonable. The primary liability, both criminally and civil for breach of the building regulations, rests on the contractor, and he will be liable to the employer if he fails to comply with the regulations. He can therefore only escape from his liability if JCT 80 clause 6.1.5 is reasonable. The contractor may have recourse in tort for an indemnity for his loss from the architect, both at common law and under the Civil Liability (Contribution) Act 1978 or for a contribution under that Act.

Quite apart from legal considerations, this must be common sense. The current building regulations regarding drains, for example, provide that they must have joints formed 'in a manner appropriate to the materials' and there is a performance specification that they must remain 'air-tight and not cause electrolytic action'. It would be quite unreasonable if a contractor were able to escape liability to the employer for defective drainage under the plea 'I only did what the contract documents required'.

4.09 Implied terms: Building Regulations

In *Street and Ano.* v. *Sibbabridge Ltd and Ano.* (1980) Judge Fay QC, sitting as a deputy High Court judge, held that it was an implied term of all building contracts that the contractor would comply with the Building Regulations. This obligation overrode the contractor's obligation under the contract to comply with the architect's design and instructions.

In that case, the defendant contractor had complied exactly with the designer's instructions as to the depth of the foundations for a garage. But due to the presence of trees in the vicinity, the foundations proved inadequate and therefore in breach of the building regulations. The judge said:

> 'The drawing provided for foundations at a depth of 2 ft 9 ins, a concrete strip under the walls and that is what [the builder] put in.
>
> It is said that there were implied terms in the contract: firstly that [the builder] would construct the garage in accordance with the building regulations which provide that "the foundation of the building shall be taken down to such a depth . . . and so constructed as to safeguard the building against damage by swelling or shrinking of the sub-soil".
>
> And secondly, "that the [builders] would use reasonable care and skill in constructing the said foundation".

These are almost common form implied terms of building contracts and I find these implied terms to be present in this case.

The term that the building would be constructed in accordance with the building regulations must flow from the fact that the builder is the person upon whom the regulations are binding. It is he who commits a criminal offence if the regulations are breached.

I have had an interesting argument addressed to me by [counsel for the builders] upon the curious situation which arises where there is an express term conveyed by the incorporation, as it undoubedly was into this contract, of the drawings . . . that had a foundation 2 ft 9 ins deep . . . contrasted with an implied term which . . . would require another type of foundation, either deeper or different in character.

In the circumstances of this case, the express term cannot be said to prevail over the implied term in any sense.

The obligation to comply with the regulations is in terms absolute. The implied undertaking to comply with the regulations must override any matter in the plans incorporated into the contract which it conflicts with.

That is only common sense, and there is evidence that it is also common practice, because the architect called by the [builder] said in his report: "It is established convention in the building industry that drawings of foundations are provisional only and are subject to review on excavation and to remeasurement on completion."

If there is such a conflict, then the implied term must prevail.'

This case is important. It makes clear that implied terms which in the words of section 7 of the Unfair Contract Terms Act arise 'by implication of law' override express terms of the contract to the contrary. By contrast, terms which might be implied to give commercial effectiveness to a contract under *The Moorcock* doctrine cannot override the contract if there is an express term which covers the same subject matter: *Les Affréteurs Réunis* v. *Leopold Walford* (1919).

However, many people, including lawyers and some judges, still believe that there cannot be an implied term if there is an express term covering the same subject matter. That is a fallacy in the case of implied terms 'arising by implication of law'. In *Young and Marten* v. *McManus Childs Ltd* (1969) the roofing sub-contractor had complied exactly with the contractual specification; nevertheless the House of Lords held he was liable for breach of a term implied by operation of law.

In the *Street* case, the judge also held that there was an implied term that the builder 'would use reasonable care and skill in constructing the foundations'.

> 'Although he honestly believed that his foundations were good ones, he had followed the plans, the foundations had been approved by the building inspector and were well constructed, he departed from the duty that the law lays upon him . . . by failing to measure up to the state of knowledge which a builder ought to have.'

However, the present author was criticised in a review of the first edition of this book for advancing this 'on the basis only of an unreported case'. But law reporters do not make the law; the judges do.

4.10 Rates and construction sites

Clause 6.2 entitles a contractor to add 'rates and other taxes' to the contract sum unless they arise from the work of a local authority or statutory undertaker contracting as a nominated sub-contractor or nominated supplier or are already priced in the bill or by way of a provisional sum.

Normally building sites are not rateable: *London County Council* v. *Wilkins*(1957). But builders' huts and cabins and the like may have a sufficient degree of permanence to be rateable. The test was set out in *Ravenseft Properties Ltd* v. *London Borough of Newham* (1976).

4.11 Person-in-charge and clerk of works

Clause 10 reads:

> 'The Contractor shall constantly keep upon the Works a competent person-in-charge and any instructions given to him by the Architect or directions given to him by the clerk of works in accordance with clause 12 shall be deemed to have been issued to the Contractor.'

There is no requirement that such a person shall be named or otherwise nominated. The expression 'foreman-in-charge' of the previous contracts has been changed, says the *JCT Guide*, to this 'the more neutral phrase'.

The powers of a clerk of works are set out in clause 12. He can only give such instructions to a contractor as an architect could under the terms of this contract and such instructions are invalid unless confirmed within two working days by the architect.

But it is noteworthy that such confirmation is only operative as an architect's instruction from the date of confirmation — not from the date when the instruction was given. (The same, of course, applies to oral instructions given by the architect himself under clause 4.3.2.1 — see [3.13].)

The contract leaves uncertain what the position is if the contractor complies with instructions given by the clerk of works, at expense to himself, but the architect fails to confirm them. It may be presumed that the contractor has to bear the expense himself. Many contractors therefore decide to ignore them.

The clerk of works is the employee of the employer, who may be vicariously liable for his negligence, with the result that damages against an architect or contractor may be reduced if the clerk of works fails to discover defects which he ought to have done: *Kensington and Chelsea Area Health Authority* v. *Adam Holden and Ors* (1984).

4.12 To obey architects' instructions

The contractor has an obligation under clause 4.1.1 to obey all instructions given by the architect which are within his powers under the contract. The contractor is entitled to be provided with details in writing under which clause of the contract the instruction is given: clause 4.2.

Disputes as to whether an instruction is empowered by the contract can be resolved by arbitration before practical completion: Article 5.2.2. If not taken to arbitration, the instruction is 'deemed for all the purposes of this contract to have been empowered by the provision specified by the architect'. This means that the contractor who does not challenge such instructions by going to arbitration will be estopped subsequently from disputing that the architect's instruction was validly given under the clause specified by the architect.

Clause 4.1.1 gives the contractor the right to object to instructions which amount to a variation under 13.1.20. This is a new addition to the contractor's rights. It follows on an enlarged definition of the word variation [3.11] which is contained in clause 13.1.2, as including alterations to obligations or restrictions contained in the contract bills.

Not only can the contractor object to any variation orders which come within clause 13.1.2; he can refuse to obey them if he makes 'reasonable objection in writing'. Again, there is provision for early arbitration contained in Article 5.2.2 but it must be virtually impossible for any arbitrator to be able to resolve such a dispute since there are no guidelines for him to follow as to what is 'reasonable objection'. He is apparently left to such native instincts as he may possess.

The only express sanction for non-compliance with the architect's instructions provided by the contract is in JCT 80 clause 4.1.2 where, if the contractor has not complied within seven days, other persons may be employed to do the work and the cost deducted from monies due to the contractor. The difficulty about this clause is that, if the contractor ignores an architect's instruction and no steps are taken to employ any other person within a reasonable period, the employer may be taken to have waived his rights conferred by this clause. If the architect decides not to press disregard of his instructions to arbitration, he should write to the contractor reserving his employer's right to proceed for breach of contract.

4.13 Setting out

The obligation is on the contractor to set out accurately the work in accordance with the architect's instructions: JCT 80 clause 7. This repeats word for word the old JCT 63 clause, including the incomprehensible words 'unless the architect shall otherwise instruct, in which case the contract sum shall be adjusted accordingly'. Does this mean that the architect can instruct somebody else to set out the works and deduct an appropriate amount from the contract sum? Does it mean that the architect can, if the contractor sets them out inaccurately, not require him to amend any errors at his own cost but call in others to do that and adjust the contract sum accordingly? Or does it mean that the architect can exempt the contractor from the obligation to remedy and make an appropriate reduction in the contract sum? If so, how can this be assessed?

4.14 Prohibition against assignment

JCT 80 contains the familiar prohibition, now in clause 19, against assignment of the contract. The wording has been altered to prohibit both parties; formerly only the contractor was prohibited from assignment:

> 19.1 'Neither the Employer nor the Contractor shall, without the written consent of the other, assign this Contract.'

The words, as they stand, are misleading. Nobody can ever assign a contract; all they can ever do is to assign the benefits of a contract — never the obligations. The assignment of the benefits of a contract is the assignment of a chose-in-action and is governed by section 136 of the Law of Property Act 1925, to which reference will be made later in this paragraph.

The real meaning of these words, in the light of their ancestry, is that they are intended as a prohibition against vicarious performance of the obligations of the contract and must be so construed. Originally 'neither at law nor in equity could the burden of a contract be shifted off the shoulders of a contractor on to those of another, without the consent of the contractee'. That flowed from the essential nature of the action of *assumpsit*, the original method of enforcing contracts not under seal. Later, the common law would allow vicarious performance, that is performance by another person, without the consent of the other party, but only where the obligations were not personal. Even where clothes are sent to cleaners, the Court of Appeal held in *Davies* v. *Collins* (1945) that this was a personal obligation which could not be sub-contracted out without it being a breach of contract. But the repair of railway wagons, with 'a rough description of work which ordinary workmen conversant with the business would be perfectly able to execute', was held to be one that could adequately be performed by somebody other than the party who contracted to do it: *British Wagon Co.* v. *Lea* (1880).

Generally speaking in most building work, in the absence of custom or contract to the contrary, there cannot be vicarious performance. The common practice of sub-contractors sub-sub-contracting out part of their work without consent is normally a breach of their contract. It is this which it is intended to prohibit, even though clause 19.2 also deals with the same subject matter:

> 'The Contractor shall not without the written consent of the Architect . . . (which consent shall not be unreasonably withheld) sub-let any portion of the Works.'

Where the architect does approve, the sub-contractor becomes a domestic one. The form intended to be used for the contractor/sub-contractor relationship is DOM/1, as it is with a clause 19 sub-contractor.

Assignment of monies due under the contract

So far as monies due to the contractor are concerned, section 136 of the Law of Property Act 1925 reads in part:

> 'Any absolute assignment by writing under the hand of the assignor . . . of any debt or other legal thing in action . . . is effectual in law to pass and transfer . . .:
> (i) the legal right to such a debt or thing in action
> (ii) all legal and other remedies for the same
> (iii) the power to give a good discharge for the same without the concurrence of the assignor. . . .'

For a legal assignment under this section, consent of the debtor is not required.
All that is necessary is:

(1) it must be in writing
(2) it must be signed by the assignor (i.e. the contractor)
(3) it must be absolute and not by way of charge
(4) the debtor i.e. the employer must be given notice of the assignment and the assignment is operative immediately on 'the date of the notice'.

Consideration between assignor and assignee is not necessary. Although the section refers solely to legal choses in action, the courts have held that it is equally applicable to equitable ones such as an interest in a trust fund.

Even where section 136 is not complied with, the courts will enforce equitable choses in action, subject to the requirements of section 53(1)(c) of the Law of Property Act 1925:

> 'A disposition of an equitable interest or trust subsisting at the time of the disposition must be in writing signed by the person disposing of the same.'

Notice to the debtor is not essential to an equitable assignment and the assignment can be by way of charge. There can therefore be (a) a legal assignment of a legal chose in action, (b) a legal assignment of an equitable chose in action, (c) an equitable assignment of a chose in action and (d) an equitable assignment of an equitable chose in action.

However, in *Helstan Securities Ltd* v. *Hertfordshire County Council* (1978), the court was concerned with the ICE Conditions, 4th Edition, which provided:

> 'Condition 3 The contractor shall not assign the contract nor any part thereof *or any benefit or interest therein or thereunder* without the written consent of the employer' [author's emphasis].

It will be seen that the wording is considerably more emphatic and detailed than JCT 80 clause 19.1.

The contractors were a company called Renhold who assigned monies due from Hertfordshire County Council to the plaintiffs, Helstan Securities Ltd.

Mr. Justice Croom-Johnson held that, in view of the terms of the contract, there had been no valid assignment. He said:

> 'The clause is obviously there to let the employer retain control of who does the work. Condition 4, which deals with sub-letting, has the same object.
>
> Closely associated with the right to control who does the work, is the right at the end of the day to balance claims for money due on the one hand against counterclaims, for example, for bad workmanship on the other.
>
> The plaintiffs say that such a counterclaim may be made against the assignees instead of against the assignors. But the debtor may only use it as shield by way of set-off and cannot enforce it against the assignees if it is greater than the amount of the debt: *Young* v. *Kitchin* (1878).
>
> Why should they have to make it against people whom they may not want to make it against, in circumstances not of their choosing, when they have contracted that they shall not?'

That judgment appears to be based upon several fallacies. The debtor was the County Council; the assignor was Renhold; the assignee was Helstan Securities.

Every assignee takes subject to the equities existing between the debtor and the assignor: so that the County Council was entitled to withhold monies due to the contractor in respect of counterclaims or by way of set-off. The assignee can never be in a better position than the assignor.

Moreover, assignment of a chose in action, such as money due, in no way discharges the contractual obligations of the assignor. It is he and not the assignee, who remains liable on the obligations of the contract, so that if the Council's claim exceeds the amount of the debt, they could still sue the contractors.

Possibly, one contracting party can validly undertake to the other not to exercise his rights under the Law of Property Act 1925 and in equity. But it is submitted that that can only be done expressly and it does not seem to have been achieved by the vague wording of JCT 80 clause 19.1.

4.15 Indemnities given by the contractor

The contractor provides indemnities to the employer against statutory charges. For the duration of such indemnities see [2.12].

Clause 6.2

From this it would appear that any fees payable in respect of planning applications after the contract is entered into and those in respect of building inspection are to be paid by the contractor. But since the section also provides that such fees or charges are to be added to the contract sum, this appears to be no more than circuitous — unless they are caught by the proviso that such fees are 'in respect of work executed . . . by a local authority'. The words are ambiguous but 'executed' is inept to describe either planning or building control fees, and therefore relates only to work which a local authority may perform or a nominated sub-contractor.

Royalties: clause 9.1

'All royalties or other sums payable in respect of the supply and use in carrying out the Works as described by or referred to in the Contract Bills of any patented articles, processes or inventions shall be deemed to have been included in the Contract Sum and the Contractor shall indemnify the Employer from and against all claims, proceedings, damage, costs and expense which may be brought or made against the Employer or to which he may be put by reason of the Contractor infringing or being held to have infringed any patent rights in relation to any such articles, processes or inventions.'

Injury to persons or damage to property: clause 20.1

'The Contractor shall be liable for, and shall indemnify the Employer against, any expense, liability, loss, claim or proceedings whatsoever arising under any statute or at common law in respect of personal injury to or the death of any person whomsoever arising out of or in the course of or caused by the carrying out of the Works, unless due to any act or neglect of the Employer or of any person for whom the Employer is responsible.'

Clause 20.2

'Except for such loss or damage as is at the sole risk of the Employer under clause 22B or 22C (if applicable) the Contractor shall be liable for, and shall indemnify the Employer against, any expense, liability, loss, claim or proceedings in respect of any injury or damage whatsoever to any property real or personal in so far as such injury or damage arises out of or in the course of or by reason of the carrying out of the Works, and provided always that the same is due to any negligence, omission or default of the Contractor, his servants or agents or of any sub-contractor, his servants or agents' [8.02].

4.16 Access to the works and premises

Strangely enough it was thought necessary to provide in JCT 80 that the architect and his representative should have access to the works at all reasonable times: clause 11.

But the clause, which reproduces clause 9 of JCT 63, also provides that the architect shall have access to the workshops or other places of the contractor where work is being prepared for the contract. That, certainly, is a right that without this clause the architect would not have. Since, because of privity of contract, the contractor cannot by JCT 80 confer similar rights on the architect in respect of sub-contractors' premises, the clause imposes an obligation on the contractor to write in similar terms into any sub-contract into which he enters, and further an obligation that the contractor 'shall do all things reasonably necessary' to make such right effective. Presumably this is wide enough to require the contractor to institute proceedings against a sub-contractor if the architect or his representatives are refused admission to sub-contractors' premises.

Curiously, there is no similar provision where one might expect to find it in NSC/2 or NSC/2a, the two employer/nominated sub-contractor contracts [5.06]. The result is that the architect will have no right whatsoever to have access without consent to the premises of a sub-contractor who has been given a 'preliminary notice of intention to nominate' and who has been called upon to design and/or fabricate in advance of nomination some sub-contract works. Once the sub-contractor has been nominated by NSC/3, and when he has entered into NSC/4 with the main contractor, the architect's right to have access to his works will only be enforceable by the contractor. This will be subject to the curiously worded NSC/4, clause 25:

> 'The contractor and the architect and all persons duly authorised by either of them shall at all reasonable times have access to *any work* [author's italics] which is being prepared for or will be utilised in the sub-contract works unless the architect shall certify in writing that the sub-contractor has reasonable grounds for refusing such access.'

This is rather less than the unqualified right of access to the sub-contractor's premises which the contractor, by JCT 80 clause 11, undertakes to secure for the architect. Access to work is not the same as access to 'workshops and places'.

The draftsmen of the JCT 80 apparently take the view that the contractor is not merely a licensee on the employer's premises but is in possession of them because they created in clause 29 an obligation 'to permit the execution of work not forming part of this contract': clause 29.1.

That sub-clause only operates when the contract bills have indicated that such work is to be done.

Where the contract bills do not provide such information, it would appear that the contractor is entitled to refuse his consent to other persons entering on the site to execute work, except in so far that 'such consent shall not be unreasonably withheld': clause 29.2.

The position regarding statutory undertakers in connection with this and other clauses is discussed in [5.16].

Clause 29.3 seems to cloud the issue considerably:

'Every person employed or otherwise engaged by the Employer as referred to in clauses 29.1 and 29.2 shall for the purpose of clause 20 be deemed to be a person for whom the Employer is responsible and not to be a sub-contractor.'

Clause 20 is the clause that gives the employer an indemnity against proceedings arising out of personal injuries or death 'unless due to any act or neglect of the employer or any person for whom the Employer is responsible'.

Clause 29.3 refers to such persons being 'deemed to be'. Normally those words would mean that such a person is to be considered as one for whom the employer is responsible even if it is manifest that he is not. The concluding words 'deemed . . . not to be a sub-contractor' on the ordinary interpretation would suggest that even if the person concerned is in fact a sub-contractor, he is to be treated as if he were not.

At the same time, it should be stressed that this position applies only 'for the purpose of Clause 20' and does not apply to other clauses under the contract. It may, however, considerably reduce the value of the contractor's indemnity to the employer if he is to be held responsible for whose who are in fact sub-contractors to the main contractor but, because of this clause, are deemed not to be.

4.17 Antiquities

The contractor is under an obligation to report to the architect or the clerk of the works 'all fossils, antiquities and other objects of interest or value which may be found upon the site': JCT 80 clause 34. He is also obliged not to disturb them and to take necessary steps to preserve them, including ceasing work. In return, he is entitled to any loss and/or expense incurred by complying with the instruction which the architect is obliged to issue. His loss and/or expense claim can arise from suspension of the works or from complying with the architect's instructions.

4.18 Contractor assumed to be a limited company

The Standard Form is drafted with the assumption that the contractor is a limited liability company.

Attention is drawn in a footnote to JCT 80 clause 35.13.5.4.4 to the fact that that clause needs to be amended if the contractor is a partnership or an individual. The employer has a right, created by clause 35.13.5.3, to pay a nominated sub-contractor directly if the contractor has not paid any sums previously certified as due to the sub-contractor. This right is made to cease in precisely those circumstances where it would be of most value to a sub-contractor, namely if the contractor company has a petition presented against it for winding up or the shareholders pass a valid resolution for voluntary liquidation.

Presumably this is because the JCT was under the impression that the decision of the House of Lords in *British Eagle International Air Lines* v. *Compagnie Nationale Air France* (1975), on very different facts, somehow overruled the Court of Appeal decision in *In re Tout and Finch* (1954) which specifically dealt with employers' right under the JCT contract to pay a sub-contractor direct.

In this respect, the JCT 80 form has been changed from JCT 63 to the disadvantage of a nominated sub-contractor.

How 35.13.5.4.4 should be amended is not suggested by the authors of JCT 80 and there seems to be no practical way it can be done simply. The presentation of a petition to the court for compulsory winding-up or a resolution to wind up are single, ascertainable events which have to be advertised in the *London Gazette*. The liquidator of a company has rights which relate back to either event. On the other hand, the title of a trustee in bankruptcy of a partnership or an individual relates back not to the date of the bankruptcy petition but to the first available 'act of bankruptcy' up to three months before the petition. To amend 35.13.5.4.4 to make the operative date which triggers off the last of the employer's right to pay a nominated sub-contractor directly the date of the presentation of a bankruptcy petition would not have the same legal effect at all as that set out in the sub-clause as applicable to companies.

There are other provisions in the Standard Form which make it unsuitable for use by contractors who are partnerships or individuals, but it is not thought necessary to elaborate on them since it will be rare that the form will be used except where the contractor is a company. Architects should, however, bear this in mind when advising their employers which contract to use if a proposed contractor is a private individual.

Chapter 5

Sub-contractors, suppliers and statutory authorities

5.01 Types of nominated sub-contractors

As a result of what were thought to be defects in the JCT 63 contract, exposed by the three House of Lords' cases of *Gloucestershire County Council* v. *Richardson* (1969), *Bickerton* v. *North West Metropolitan Hospital Board* (1969) and *Westminster City Corporation* v. *Jarvis* (1970), JCT 80 has made substantial alterations to the procedure for nominating sub-contractors.

In addition for the first time in these contracts, it is recognised that modern buildings are so complex and contain such highly specialised services, that it is inevitable that certain parts must be designed and constructed by specialists.

The architect is no longer in a position to design the whole of any required building, nor has he been for many years. In spite of that, he cannot abdicate responsibility by assigning to a specialist sub-contractor the design: *Moresk Cleaners* v. *Hicks* (1966). And even where the employer himself negotiates directly with a sub-contractor and gives his architect express directions as to whom to nominate, the architect will be liable if details of the sub-contractor's design, which he has had and approved, prove defective: *Holland Hannen & Cubitt* v. *Welsh Health Technical Services Organisation* (1981).

JCT 63 attempted to protect the contractor's position to some extent by granting him exemption from liability to pay liquidated damages to the employer if progress of his work was retarded 'by delay on the part of a nominated sub-contractor'. Litigation inevitably resulted from that singularly inept wording (repeated in JCT 80 clause 25). Clearly what the JCT wanted the contractor to obtain was an extension of time if the work of the contractor was delayed by failure on the part of a nominated sub-contractor to perform his contract in any way. Those words do not achieve that objective.

But the net effect of these contractual provisions was that the contractor could obtain an extension of time for 'delay on the part of a nominated sub-contractor', resulting in the employer losing his liquidated damages, without the employer having any recourse against the sub-contractor he nominated.

Further, if the nominated sub-contractor's design input, materials or workmanship were defective, the employer had no recourse in contract against his nominee since there was no contractual relationship between them. Moreover, in the state of the law at that time, it appeared that the contractor was not liable either — apart from the defects liability period.

There are now no less than five methods whereby a sub-contractor can become a nominated one under JCT 80:

(1) the 'basic method' [5.03]
(2) the 'alternative method' [5.05]
(3) other nominations [5.08]
(4) 'substituted nominations' under clause 35.24 [5.09]
(5) re-nominations [5.10].

The procedure whereby an architect negotiates with a sub-contractor, concludes terms with him without consultation with the main contractor, and then makes a nomination of him as the person with whom the contractor is required to enter into contractual relations, developed early in the last century. It was held in *Leslie* v. *Metropolitan Asylums* (1901) and *Mitchell* v. *Guildford Union* (1903) that the architect in such circumstances was not acting as the agent of the employer and therefore the employer was not liable to the contractor for defective performance by the sub-contractor.

Nomination seemed to give the employer the best of both worlds. He could select his own specialist sub-contractor, at his own price, and accept no responsibility whatsoever for the performance of his nominee, but there were substantial drawbacks.

5.02 Contractual relationships under JCT 63

Contractual relations under JCT 63 can be illustrated graphically (Figure 2).

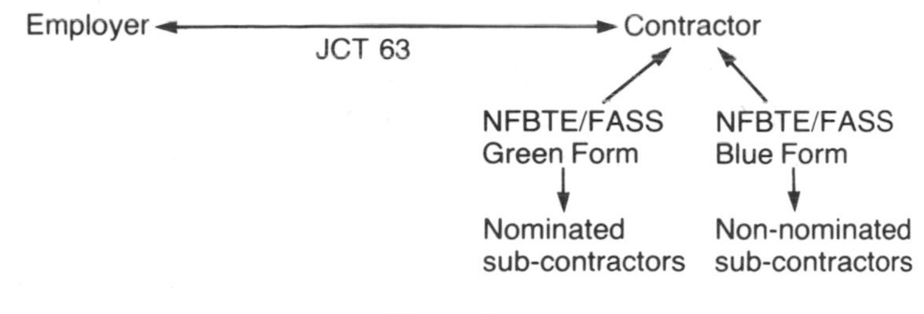

Figure 2

Normally the first contract to come in to existence was that between the employer and his architect, which might or might not have been in the standard RIBA Conditions of Engagement, which were then in use. Following that, the main contract between the employer and the contractor under JCT 63 was placed. Thereafter, the contractor placed contracts on the NFBTE/FASS Green Form with nominated sub-contractors, and on the NFBTE/FASS Blue Form or his own form with other sub-contractors, termed by JCT 80 'domestic sub-contractors'. The position therefore was as shown in Figure 2.

One result of this arrangement was that there resulted no contractual relationship between the employer and any sub-contractor. On the principles outlined above, the employer could not sue a sub-contractor in contract, however badly the work was done or whatever delay was caused. By virtue of clause 23(g) of JCT 63, a main contractor could escape from an obligation to pay liquidated damages to the employer if he secured a certificate from the architect for an extension of time caused 'by delay on the part of nominated contractors . . . .' The employer could lose his right to liquidated damages and have no remedy in contract against the sub-contractor at fault.

This situation was considered in the Court of Appeal in *Jarvis* v. *Westminster Corporation* (1969). There, Lord Justice Salmon (as he then was) said: 'Clause 23(g) is highly anomalous . . . it is in my view unjust and absurd.'

The same case went to the House of Lords and there Lord Wilberforce said:

> 'I cannot believe that the professional body, realising how defective this clause is, will allow it to remain in its present form . . . it leaves the employers to bear the loss caused by a delay for which they are in no way to blame and allows the party at fault . . . to escape from the liability they otherwise would have to bear.'

As a result, the House of Lords held that, where a nominated sub-contractor does defective work and then withdraws from the site, delay caused in consequence to the contractor is not 'delay on the part of a nominated sub-contractor' [7.11].

The 1980 edition retains the same wording of clause 23(g) in what is now clause 25.4.7.

After the case of *Westminster City Corporation* v. *Jarvis* (1970) in the House of Lords the RIBA issued a 'Form of Warranty' whereby a nominated sub-contractor, in consideration of nomination, gave a warranty to the employer that he would give no occasion to the contractor to obtain an extension of time.

If, but only if, this form was used, the position then became as shown in Figure 3.

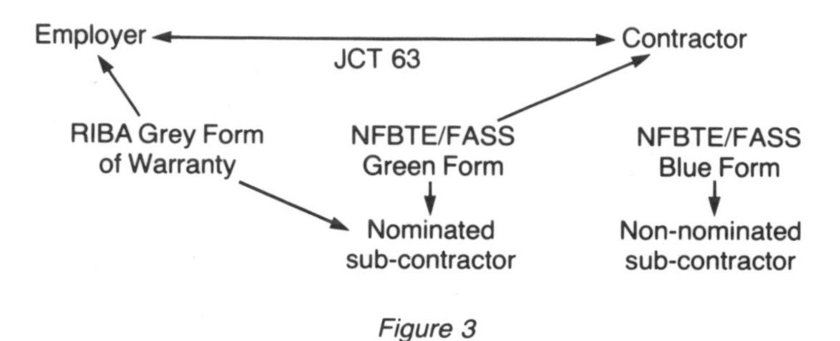

Figure 3

To deal with these problems, the 'basic' and 'alternative' methods of nomination have been introduced.

5.03 The 'basic method' of nomination under JCT 80

The scheme envisaged as the 'basic method' of nomination under JCT 80 can be diagrammatically represented — see Figure 4.

The procedure where the basic method is to be used is as follows:

Stage 1 The architect completes those relevant parts of NSC/1 with a top copy and *two* carbons.

The architect then sends these three pieces to sub-contractors, together with NSC/2; the architect must indicate whether NSC/2 is to be signed or under seal.

Stage 2 A sub-contractor completes all three NSC/1 copies and signs each.

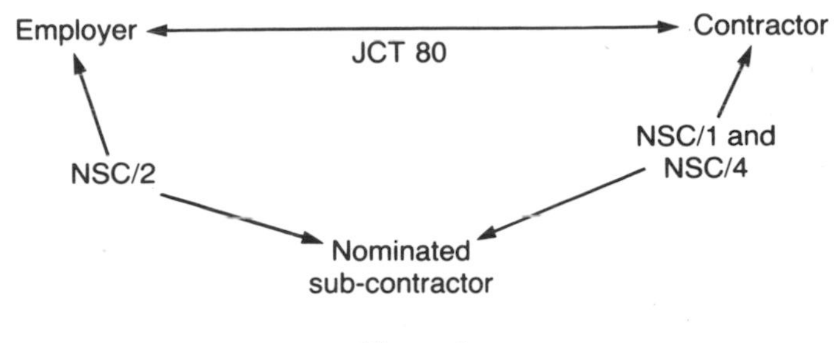

Figure 4

He signs (or seals) NSC/2.
He then returns all the documents to the architect.

Stage 3 The architect having selected one sub-contractor signs NSC/1 on page 1 under the sub-contractor's signature as 'Approved by the Architect on behalf of the Employer'.

The architect then has to procure the signature or seal of the employer to NSC/2.

The employer is to retain NSC/2 but to return to the architect a certified true copy.

Stage 4 The architect sends to the sub-contractor a certified true copy of NSC/2 signed or sealed by the employer.

Stage 5 The architect then has to send to the main contractor: preliminary notice of nomination of sub-contractor (see JCT 80 clause 35.7.1);

NSC/1 completed by sub-contractor (all three copies);

A copy of NSC/2 as completed by the employer and the sub-contractor.

Stage 6 The main contractor has to complete the rest of NSC/1 in agreement with the sub-contractor; that is, Schedule 2 'Particular Conditions', and sign all three copies.

He returns all three copies to the architect.

Stage 7 The architect now issues his formal notice to the contractor nominating the sub-contractor on NSC/3. With it goes the original completed NSC/1.

Stage 8 The architect sends a copy of NSC/3 to the proposed sub-contractor; with it goes a certified true copy of NSC/1.

Stage 9 The contractor and nominated sub-contractor complete NSC/4.

5.04 Period of validity of tender

It will be noted that in Stipulation 4 it is intended that the sub-contractor himself should insert a period for which his tender will be valid. The result may be that the architect will receive a number of different tenders valid for different periods and comparison between them may be difficult. It is suggested therefore that the architect should specify this period.

5.05 The 'alternative method'

Where the architect does not propose to use the tender form NSC/1, and uses some other method of selecting a nominated sub-contractor, then the appropriate employer/sub-contractor form is NSC/2a and the appropriate contractor/sub-contractor form is NSC/4a.

These two methods are the only ones that establish contractual relationships between the employer and the sub-contractor and it is therefore proposed to look in outline at the terms of those contracts.

5.06 Terms of NSC/2 or NSC/2a

The key to both these documents is contained in NSC/2 clause 3.4 and NSC/2a clause 2.4 which contain the same words:

> 'The Sub-Contractor shall so perform the Sub-Contract that the Contractor will not become entitled to an extension of time for completion of the Main Contract Works by reason of the Relevant Event in clause 25.4.7 of the Main Contract Conditions.'

This is an express warranty that the sub-contractor will not give occasion to the contractor to obtain an extension of time for 'default on the part of a nominated sub-contractor'. If the sub-contractor in breach of this clause does so, he will be liable to the employer for the liquidated damages that the employer has lost by reason of the extension of time. It is for this reason that NSC/1 in Schedule 1, paragraph

10 specifies the liquidated and ascertained damages under the date for completion of the main contract. The sub-contractor cannot be heard to say in consequence that he did not know what loss of liquidated damages the employer will suffer if he causes the employer to lose them.

Under the alternative method, the sub-contractor has no formal notice of the amount of liquidated damages. In his preliminary negotiations with him, the architect should take care to advise the sub-contractor of the completion date of the main contract and the amount of liquidated damages. Otherwise, the sub-contractor may be able to raise the plea that he had no notice at the time when he entered into the contract of the amount of the possible damages that may be caused to the employer by breach of this warranty under the rule in *Hadley* v. *Baxendale* (1854).

The sub-contractor who enters into NSC/2 or NSC/2a contracts with the employer gives an express warranty that 'he has exercised and will exercise skill and care' in (a) the design, (b) the selection of material and goods and (c) the satisfaction of any performance specifications. This is, of course, an inadequate and worthless promise, being merely an attempt to apply the *Bolam* standard of reasonable care. There can be no doubt that the law will import into this contract, as into all construction contracts [4.03], implied terms that the design will be fit for the purpose required, that any performance specification will be complied with *exactly* and that all materials and goods will correspond with the description, will be of mercantile quality and reasonably fit for the purpose required. In spite of this wording, sub-contractors will be required to guarantee their work; and rightly so.

In addition, there will be express statutory warranties now contained in the Supply of Goods and Services Act 1982.

5.07 Design and fabrication before nomination

Sub-contractors who have entered into a contract with the employer on NSC/2 or 2a undertake even before they receive nomination to proceed with (a) the design, (b) the ordering, and (c) the fabrication of the 'material and goods if so required by the architect'.

If they do not receive nomination, they are entitled only 'to any expense reasonably and properly incurred by the sub-contractor in carrying out work in the designing of the sub-contract works' and for 'any materials and goods properly ordered by the sub-contractor for the sub-contract works'.

It is surprising that the representatives of specialist sub-contractors have agreed to these terms. For they appear to afford to the sub-contractor who has entered into NSC/2 or 2a and received orders from the architect to design and fabricate the sub-contract works the bare expense thereof without provision for any profit margin.

Moreover, on payment of such 'expenses' the employer may use the design 'for the purposes of the sub-contract works but not further or otherwise' and 'any materials and goods so paid for shall become the property of the employer'.

The way appears open, therefore, for an architect to order a specialist sub-contractor who has signed NSC/2 or 2a to design and fabricate sub-contract works, to refuse the sub-contractor nomination and then to acquire the designed and manufactured works at bare cost.

Moreover, it would appear that if these provisions are implemented, it will no longer be a 'work and materials' contract and the sub-contractor will become merely a seller of goods within the meaning of the Sale of Goods Act 1979 and subject to all the provisions of that Act, including the implied terms and provisions.

The copyright, trade mark and patent law aspects of design appear to be dealt with inadequately by the provision in NSC/2 clause 2.2.3 that the employer may use the design of the sub-contractor for the sub-contract works (or NSC/2a clause 1.2.3). It may well be that these clauses amount to a valid but restricted licence, but it by no means follows that proprietary rights can be dealt with in such a cavalier fashion.

5.08 Other nominations

As has been pointed out earlier [5.01] a sub-contractor can become a 'nominated sub-contractor' for the purposes of JCT 80 clause 35, even though neither the 'basic' nor the 'alternative' method of nomination is used, and without entering into a contract on NSC/2 or 2a with the employer.

If this happens, an architect should be alert to the possibility that such a nominated sub-contractor will have no contractual relationship whatsoever with the employer since neither NSC/2 nor 2a will have been used. In particular, the employer may have no recourse in contract against such a sub-contractor if the employer loses his right to liquidated damages, because of 'delay on the part of a nominated sub-contractor'. An architect should therefore seek, before nominating such a sub-contractor, an undertaking that in consideration of nomination as in NSC/2 clause 3.4:

> 'The Sub-Contractor shall so perform the Sub-Contract, that the Contractor shall not become entitled to an extension of time . . .'

under JCT 80 clause 25.4.2.

He should also advise the sub-contractor of the date for completion of the main contract and also the quantum of liquidated damages.

5.09 Substituted nominations

JCT 80 attempts to deal with the criticisms made of JCT 63 that it made no provisions for what was to happen if a nominated sub-contractor ceased to perform his obligations.

Clause 35.24 deals with substituted nominated sub-contractors. It is restricted to the three cases:

- Where the architect is of the opinion that the sub-contractor is in default so that his employment can be determined under clause 29 of NSC/4 or 4a. These clauses contain provisions similar to the provisions for determination of a main contractor's employment which are contained in JCT 80 clause 27 [9.02].
- Where the sub-contractor becomes insolvent.
- Where the sub-contractor determines his employment validly under clause 30 of NSC/4 or 4a, similar in terms to JCT 80 clause 28 [9.06].

JCT 80 clause 35.24.4 then specifies what the architect shall do in any of these circumstances, with the ultimate obligation that 'the architect shall make such further nomination of a sub-contractor . . . as may be necessary'.

There then follows a provision, in clause 35.24.6:

> 'Where clause 35.24.3 applies the Architect shall make such further nomination of a sub-contractor in accordance with clause

> 35 as may be necessary to supply and fix the materials or goods or to execute the work but any extra amount over the price of the Nominated Sub-Contractor who has validly determined his employment under his Sub-Contract, resulting from such further nomination may be deducted by the Employer from monies due or to become due to the Contractor under this Contract or may be recoverable from the Contractor by the Employer as a debt.'

It will be noted that this clause, which requires a contractor to pay the increased costs for a substituted sub-contractor, only applies where the sub-contractor has lawfully determined his own employment under NSC/4 or 4a clause 30. These changes only entitle the sub-contractor to withdraw if the contractor 'without reasonable cause' suspends the sub-contract works etc.

It does not apply where there is a renomination because the sub-contractor's contract has come to an end as the result of his own default or insolvency. There, presumably, although the contract is silent about it, the employer must pay. JCT 80 clause 35.24.2 applies to such circumstances.

None of these provisions applies to the situation where the sub-contractor has simply withdrawn from the site or legitimately suspended work.

5.10 Renominated sub-contractors

Apart from the provisions referred to above in clause 35, the situation may arise where a nominated sub-contractor withdraws from the work, being neither in default, nor entitled to determine his employment, nor insolvent.

In these circumstances, it is clear that the employer is under an obligation to renominate: *Bickerton* v. *North West Metropolitan Hospital Board* (1969). That is an obligation which, if not performed within a reasonable time, may entitle the contractor to damages for breach of contract, if his own work is impeded thereby, and may render time at large for the purposes of liquidated damages.

Express provision does not appear to have been made to cover the situation where a sub-contractor suspends work without determining either his employment or the contract. This, it would appear, can happen [9.01].

5.11 Contractor doing nominated sub-contract work

JCT 80 in clause 19.5.2 makes clear what was uncertain in the standard contracts before the *Bickerton* case:

> 'Contractor is not himself required to supply and fix materials or to execute work which is to be carried out by a Nominated Sub-Contractor.'

There is one exception to this however. Clause 35.2.1 allows the contractor, if he so wishes and if 'the architect is prepared to receive tenders from the contractor', to tender for items which have been set out in the Appendix to JCT 80.

The Appendix therefore contains a paragraph: 'Work reserved for Nominated Sub-Contractors for which the Contractor desires to tender.'

Where a main contractor's tender for such work is accepted, the clause then provides that he shall not sub-let the work to a domestic sub-contractor without the consent of the architect, provided that, where an item for which the architect intends to nominate a sub-contractor is included in architect's instructions issued under clause 13.3, it shall be deemed, for the purposes of clause 35.2.1, to have been included in the contract bills and the item of work to which it relates shall likewise be deemed to have been set out in the Appendix.

Clause 35.2.2 states:

> 'It shall be a condition of any tender accepted under clause 35.2 that clause 13 shall apply in respect of the items of work included in the tender as if for the reference therein to the Contract Drawings and the Contract Bills there were references to the equivalent documents included in or referred to in the tender submitted under clause 35.2.'

Clause 13 is of course the one that deals with variations. The result of this clumsy wording is that, in effect, what the architect originally contemplated was to be sub-contract work becomes part of the main contract if the contractor's tender is accepted.

One commentator has claimed that the above statement is the exact opposite of what clause 35.2.2 actually says. The reader can read it for himself. It clearly says that if the main contractor's tender for the sub-contract work is accepted, that work shall be treated as a variation to the main contract and be subject to the same principles of valuation of variations as the main contract, substituting the tender documents in respect of that work for the contract drawings and bills.

5.12 Contractor's consent required for nominations

Clause 35.4.1 provides:

> 'No person against whom the Contractor makes a reasonable objection shall be a Nominated Sub-Contractor.'

Neither JCT 80, nor its predecessors, gives any indication of what is a reasonable objection but it is generally regarded as including the financial stability of the nominated sub-contractor and his work record. But Lord Justice Sachs said in the *Bickerton* case:

> 'He has no right to object to any details in the specification in the sub-contract, any more than to the sub-contract price . . . his sole right of objection is to the person nominated.'

That observation must now be regarded as modified in the light of subsequent cases about what the industry and the contracts describe as 're-nomination' but which would more accurately be described as the nomination of replacement sub-contractors.

It is now clear that it is not only valid to object to the person nominated, but also to that person's programme and/or method or work.

In the Court of Appeal, in *Percy Bilton Ltd* v. *Greater London Council* (1981) (which subsequently went to the House of Lords), Lord Justice Stephenson said that the plaintiff argued:

> 'that at the time of the application for re-nomination the new sub-contractor's date for completion was later than the plaintiff's date for completion . . . therefore the time provision must go completely . . . I do not accept that argument. The contractor, faced with a sub-contract with such a provision . . . would be entitled to refuse to accept that sub-contractor under clause 27' [of JCT 63 — now JCT 80 35.4.1].

Lord Fraser in the House of Lords expressed the same view.

In *Fairclough Building Ltd* v. *Rhuddlan Borough Council* (1983), Judge Smout said:

> 'where there is an over-run, as in this instance, in the absence of any specific undertaking by the architect that the time limits under the main contract will be adjusted so that the main contract and the sub-contract are compatible one with the other, the main contractors are entitled on that account to reject under 27(1)(ii).'

He also held that the main contractors were entitled to reject a nomination for a replacement sub-contractor unless that sub-contractor was directed to do the whole of the original sub-contractor's work, including any necessary remedial work. The architect could not, as he had done, issue an instruction that the main contractor should 'investigate and make good' work done by the previous sub-contractor. Main contractors have neither the right nor the obligation to do any of the prime cost work themselves. The judge added:

> 'By reason of the failure of the architect to require [the replacement sub-contractor] to do the remedial as well as the completion work, I hold that the plaintiffs were entitled to reject the nomination.'

The latest date on which such objection can be made is seven days from receipt by the contractor of the nomination by the architect: JCT 80 clause 35.4.2.

5.13 Domestic sub-contractors

This term is introduced to main work and material sub-contractors who are not nominated by the architect. A domestic sub-contractor can be used in two circumstances: where the architect makes use of the new provisions contained in clause 19 [5.15]; and where the contractor has applied for and been given approval.

5.14 Contractor not liable for sub-contractor's design?

Clause 35.21 purports to exonerate the contractor for any liability for the design element in any nominated sub-contractor's work:

> 'Whether or not a Nominated Sub-Contractor is responsible to the Employer in the terms set out in clause 2 of the Agreement NSC/2 or clause 1 of the Agreement NSC/2a the Contractor shall not be responsible to the Employer in respect of any nominated sub-contract works for anything to which such terms relate. Nothing in clause 35.21 shall be construed so as to affect the obligations of the Contractor under this Contract in regard to the supply of workmanship, materials and goods.'

It remains to be seen whether these words are sufficient to displace the terms which are now normally to be implied in construction work [4.03]. A much more emphatic exemption of liability — clause 30 in the Model Conditions — was disregarded by the judge of first instance in *IBA* v. *EMI and BICC* (1980) and ignored by the House of Lords [4.03].

5.15 Non-nominated sub-contractors: clause 19 procedure

JCT 80 makes provision for the first time, in clause 19, for the employer to provide a list of 'not less than three persons' from whom the contractor can select a sub-contractor. Clause 19.3.2.1 provides for either party with the consent of the other, 'which consent shall not be unreasonably withheld', to add an additional person to the list at any time. It has therefore been suggested, with justification, that an employer should provide an initial list of at least five names, since some specified sub-contractors may not wish to tender under this clause.

The wording of clause 19.3.1 is full of ambiguities. It requires that:

> 'Where the Contract Bills provide that certain work measured or otherwise described in those Bills and priced by the Contractor must be carried out by persons named in a list in or annexed to the Contract Bills, and selected therefrom by and at the sole discretion of the Contractor the provisions of clause 19.3 shall apply in respect of that list.'

This is taken to mean that the contractor must be supplied with a list of persons so named before he himself tenders, so that his tender can include the sub-contractor he has selected from the list and that sub-contractor's price.

A sub-contractor selected under this clause by a contractor is to be a domestic sub-contractor, so that the contractor has no right to an extension of time for any delay on his part and he accepts full responsibility for the performance of the sub-contractor he has selected.

In practice in spite of the ambiguities, this appears to be a most valuable provision in the contract, so far as the employer is concerned, and it may well be that this becomes the standard method of selecting sub-contractors for work on major contracts. Experience since 1980 has shown that invariably contractors do not select the highest tender, but the most reliable sub-contractor.

5.16 Statutory undertakers

The Gas, Electricity and Water Boards and other statutory bodies, including local authorities, may be involved either in performance of their statutory obligations or as contractors. The statutory obligations of each board are different and are amended from time to time.

The important thing to note is that while they are performing their statutory obligations they do not enter into contracts.

The earliest case of this kind appears to be *Milnes* v. *Huddersfield Corporation* (1886) where it was held that there was no contractual obligation by a water company created by Act of Parliament towards those it supplied.

The same thing was said in relation to a gas company later in *Clegg Parkson and Co.* v. *Earby Gas Co.* (1896): 'the obligation of the company if any, depends on statute and not upon contract'.

So when in October 1937 the Croydon Corporation distributed to consumers, through its mains, water that was contaminated by typhoid germs and started an epidemic in which over 300 people became infected and 43 died, the Corporation succeeded in escaping liability in contract. Had it been a contract, those injured who paid water rates would have been entitled to damages under the Sale of Goods Act 1893, since water (like milk) is a chattel. Damages would have been recoverable by them whether or not the Corporation had been negligent, since the water was clearly not fit for the purpose required. But Mr Justice Stable held that the relationship was not a contractual one, but 'a relationship between two persons under which one is bound to supply water and the other, provided he has paid the equivalent rent, is entitled to receive it': *Read* v. *Croydon Corporation* (1938).

This principle was apparently regarded as so axiomatic that none of these cases appears even to have been discussed in *Willmore* v. *South Eastern Electricity Board* (1957). Mr and Mrs Willmore started in business as poultry farmers rearing chicks by infra-red heat and the South Eastern Electricity Board promised them an adequate and constant supply of electricity to maintain lamps for that purpose. This the Board failed to do, with the result that the lamps chilled, the chicks died and Mr and Mrs Willmore were ruined financially. The judge held that the representations about proper supply of electricity were not made *animo contrahiendi*, as he put it — 'with contractual intent' — and there was no contract at all between the South Eastern Electricity Board and the unfortunate consumers. 'I have come to the conclusion that the plaintiffs, having failed to prove a contract, can have no cause of action of damages for breach of contract.'

Although they may not be liable for damages for breach of contract that does not exonerate them in tort, e.g. for negligence.

JCT 80 therefore draws a distinction between statutory undertakers performing their statutory duties and those situations where they are contractors. JCT 80 clause 6.3 provides:

> 'The provisions of clauses 19 and 35 shall not apply to the execution of part of the Works by a local authority or a statutory undertaker executing such work solely in pursuance of its statutory obligations and such bodies shall not be sub-contractors within the terms of this Contract.'

The effect of the distinction will be considered again in relation to extensions of time [7.09] and claims for 'loss and/or expense' [10.04].

5.17 Nominated suppliers

JCT 80 contains, in clause 36, an elaborate and lengthy definition of what is a 'nominated supplier'. But it first is desirable to look at what is *not* a 'nominated supplier'.

This is contained in clause 36.1.2: 'Notwithstanding that the supplier has been named in the Contract Bills or that there is a sole supplier of such materials or goods,' the term is not used for those who supply goods to be fixed by the contractor 'unless such materials or goods are the subject of a prime cost sum' in the bills. In other words, if the source of the materials has been specified in the bills and has been priced by the contractor, the supplier is not a 'nominated one'.

Where a prime cost sum has been specified in the bills, the architect is under a duty to clause 36.2 to issue instructions for the purpose of nominating a supplier.

But the definition clause 36.1 also includes others as 'nominated suppliers' and contains subtle traps for an unwary architect, particularly in regard to variation orders and the expenditure of provisional sums:

36.1.1.1 'where a prime cost sum is included in the Contract Bills in respect of those materials or goods and the supplier is either named in the Contract Bills or subsequently named by the Architect in an instruction issued under clause 36.2;

1.2 where a provisional sum is included in the Contract Bills and in any instruction by the Architect in regard to the expenditure of such sum the supply of materials or goods is made the subject of a prime cost sum and the supplier is named by the Architect in that instruction or in an instruction issued under clause 36.2;

1.3 where a provisional sum is included in the Contract Bills and in any instruction by the Architect in regard to the expenditure of such a sum materials or goods are specified for which there is a sole source of supply in that there is only one supplier from whom the Contractor can obtain them, in which case the supply of materials or goods shall be made the subject of a prime cost sum in the instructions issued by the Architect in regard to the expenditure of the provisional sum and the sole supplier shall be deemed to have been nominated by the Architect;

1.4 where the Architect requires under clause 13.2, or subsequently sanctions, a Variation and specifies materials or goods for which there is a sole supplier as referred to in clause 36.1.1.3, in which case the supply of the materials or goods shall be made the subject of a prime cost sum in the instruction or written sanction issued by the Architect under clause 13.2 and the sole supplier shall be deemed to have been nominated by the Architect.'

5.18 Nominated suppliers: tender form TNS/1

Since a contractor can obtain an extension of time under JCT 80 clause 25.4.7 for 'delay on the part of nominated . . . suppliers', the provision has been made for the employer to receive tenders on TNS/1 for the supply of goods in such a fashion as to create a contractual relationship between himself and the supplier and receive a warranty that the nominated supplier will not give the contractor occasion to obtain an extension of time.

The tender form TNS/1 contains a reprint of the voluminous conditions which are contained in clause 36.4.1 to 36.4.9.

Clause 36.4 provides that the architect shall only nominate a supplier who will enter on a contract with the contractor for the supply of materials and goods on these terms (unless the contractor agrees otherwise).

In addition to those conditions, the supplier will also have to comply with all the provisions of the Sale of Goods Act 1979.

It is not proposed to rehearse all the conditions contained in clause 36.4.1 to 36.4.9. Clause 36.4.7 however, merits special attention:

> 'that the ownership of materials or goods shall pass to the Contractor upon delivery by the Nominated Supplier to or to the order of the Contractor, whether or not payment has been made in full.'

In short, the architect cannot nominate, without the consent of the contractor, a supplier who operates a retention of title clause of any kind.

Such a clause is one where, as is permitted by section 19 of the Sale of Goods Act 1979, the seller remains the owner of the goods until payment of the price for those goods is made, or in the more elaborate clauses until all the indebtedness between the parties is discharged (see Parris: *Retention of Title on the Sale of Goods*).

It can safely be concluded that numerous copies of TNS/1 will be sent out to suppliers, with the conditions printed as they are in minuscule on the reverse, and will come back with entirely contradictory sets of conditions, namely the supplier's own. The stage will be set for what lawyers refer to as 'the battle of the forms' — namely which if either of two sets of entirely incompatible conditions applies to the contract of sale, and indeed, whether there is any contract of sale of the goods. In general in England the courts incline to name the winner of the battle as the 'last past the post': in other words, if goods are accepted by buyers with a delivery note containing certain conditions imposed by the sellers, those conditions apply: *British Road Services Ltd* v. *Arthur V. Crutchley & Co. Ltd* (1968); *Butler Machine Tool Ltd* v. *Ex-Cell-O Corporation* (1979). Here is fruitful ground for litigation, when goods with a retention of title clause in the delivery note have been accepted on site without any check on the terms of delivery.

It will also be noted that JCT 80 clause 36.4.7 displaces the presumptions set out in section 18 of the Sale of Goods Act 1979 and its predecessor. The presumption in general under the Sale of Goods Act 1979 is that property — that is, ownership — in goods passes when the contract for existing and specific goods is made. But the parties can contract otherwise, as they are taken to do here. Under JCT 80 property passes with delivery.

5.19 Nominated suppliers: warranty form TNS/1

With his tender, the intended supplier is expected to send back a warranty form establishing contractual relationship between himself and the employer.

The key warranty is to be found in TNS/1, clause 1.2:

> 'We will . . . so commence and complete delivery of the supply in accordance with the arrangements in our contract of sale with the contractor that the contractor shall not become entitled to an extension of time under SFBC [i.e. JCT 80] clauses 25.4.6 or 25.4.7 of the Main Contract conditions nor become entitled to be paid for direct loss and/or expense ascertained under SFBC clause 26.1. . . .'

It is to be hoped that suppliers realise what these cryptic words mean. It means that if the contractor secures an extension of time under clause 25 and the employer thereby loses his right to recover liquidated damages for that period, the supplier will be liable for the full extent of the named sum — which may well be vastly in excess of the price of the goods.

The rest of that term of TNS/1 is mysterious. A contractor can claim and receive an extension of time under JCT 80 clause 25.4.7 for 'delay on the part of Nominated . . . Suppliers' but he is entitled to *no* 'loss and/or expense' under clause 26 for that reason.

It can only be assumed that this reference is to JCT 80 clause 26.2.1 — 'failure to receive instructions' etc. Otherwise it would appear meaningless.

The supplier's liability may be unlimited in time. For the TNS/1 clause goes on to make the supplier give an indemnity to the employer in respect of these sums. As has been noted [2.12], the time for such indemnity only begins to run when loss has been suffered. In *County and District Properties* v. *Jenner* (1974) Mr Justice Swanwick refused to hold that the supplier of glass, who had not given the express indemnity that the work and materials sub-contractors had under the Green Form, should be held liable on an implied indemnity. Since there was no indemnity, the result was that all action against the suppliers was statute barred six years from the time when they supplied the material; whereas for the four work and materials sub-contractors, who had given an express indemnity, the action was not statute barred.

The TNS/1 form invites a supplier to put his head in this noose. Admittedly there are escapes provided in that the warranty does not apply in so far as the supplier is delayed by various things set out in TNS/2 clause 1.2:

1.2.2.1 'force majeure; or
1.2 civil commotion, local combination of workmen, strike or lock-out; or
1.3 any instruction of the Architect under SFBC clause 13.2 (Variations) or clause 13.3 (provisional sums); or
1.4 failure of the Architect to supply to us within due time any necessary information for which we have specifically applied in writing on a date which was neither unreasonably distant from nor unreasonably close to the date on which it was necessary for us to receive the same.'

Even so, most well informed suppliers will no doubt regard these as most onerous obligations for very little reward, particularly as suppliers are to be precluded from ensuring that they get paid by retaining their ownership until they do.

The form TNS/2 purports to limit the supplier's liability in so far as design and selection of materials is concerned and the satisfaction of performance specifications to the 'exercise of all reasonable skill and care': TNS/2 clause 1.1.

But these are sales of goods and the full provisions of Sale of Goods Act 1979 apply. In particular the following will be part of the seller's obligations:

Section 14(1): Where there is a contract for the sale of goods by description, there is an implied condition that the goods will correspond with the description.

Section 14(2): Where the seller sells goods in the course of a business, there is an implied condition that the goods supplied under the contract are of merchantable quality, except that there is no such condition —
(a) as regards defects specifically drawn to the buyer's attention before the contract is made; or
(b) if the buyer examines the goods before the contract is made, as regards defects which that examination ought to reveal.

Section 14(3): Where the seller sells goods in the course of a business and the buyer, expressly or by implications, makes known —
(a) to the seller, . . .
any particular purpose for which the goods are being bought, there is an implied condition that the goods supplied under the contract are reasonably fit for that purpose.

Contracting out of these obligations will be subject to the provisions of the Unfair Contract Terms Act 1977.

5.20 Employer's own works

One method of avoiding nomination of a sub-contractor, with all its attendant horrors, is direct employment by the employer. Clause 29 appears almost to encourage this, provided that sufficient information is given to the contractor in the bills to enable him to carry out his own works. Even where this information is not given in the bills, the contractor must not unreasonably withhold his consent for the execution of such work: clause 29.2.

There are, however, problems arising out of personal injuries claims (clause 20), extension of time (clause 25) and consequential loss and delay (clause 26).

Chapter 6

Certificates, payments, retention and fluctuations

6.01 The architect as certifier

'It is established law,' said Lord Radcliffe in the House of Lords in *R. B. Burden Ltd* v. *Swansea Corporation* (1954), 'that, in granting a final certificate under a building contract, the architect acts in an arbitral capacity.' As a result of being considered a quasi-arbitrator a certifying architect was held by the Court of Appeal in *Chambers* v. *Goldthorpe* (1901) to be immune from actions for negligence by either his employer or the contractor.

That 'established law' was swept away by the decision of the House of Lords in *Sutcliffe* v. *Thackrah* (1973). An architect had given an interim certificate for £3090 on which the builder had been paid. Part of the work included in the certificate turned out to be defective. Normally, such an overpayment could have been adjusted on subsequent certificates but in this case the builder became insolvent and the overpaid money became irrecoverable. On a preliminary point of law, it was held that an architect who was called on to certify the value of work done for an interim certificate under any building contract was liable in both contract and tort to his employer for negligent over-certification. It would therefore appear that, if he is not immune as a quasi-arbitrator, he is also liable in tort to a contractor for negligent under-certification: *Stevenson* v. *Watson* (1879).

The standard of care required from professional men is that set out in *Charlesworth on Negligence*, 6th Edition, paragraph 945:

> *Nature of Duty*
> 'The duty of architects, surveyors . . . and engineers is to use the reasonable care and skill of such persons of ordinary competence measured by the professional standard of the time.'

This is commonly known as the *Bolam* standard, from the case *Bolam* v. *Friern Hospital* (1957) in which it was formulated for the guidance of a jury. Dealing with the standard applicable to an independent valuer in *Belvedere Motors Ltd* v. *King* (1981) Mr Justice Kenneth Jones said:

> 'The defendant can only be found guilty of negligence if it can be shown that he has omitted to consider some matter which he ought to have considered, or that he has taken into account some matter which he ought not to have taken into account or in some way has failed to adopt the procedure and practices accepted as standard in his profession, and has failed to exercise the care and skill which he, on accepting the appointment, held himself out as possessing.'

It is evident therefore that the architect must be liable to his employer if he issues a certificate to the contractor for payment of on-site goods without satisfying himself that the contractor can pass such a title to the employer as will make the employer the owner of the goods for which he has to pay.

In *Ashwell Scott Ltd* v. *Scientific Control Systems Ltd; Milton Keynes Development Corporation* (third parties) (1979), it emerged in the course of the hearing that there was an ingenious arrangement between all the parties whereby the air-conditioning to be installed was leased from Eastlease Ltd, the leasing subsidiary of the Norwich Union, with the benefit of 100 per cent write-off against corporation taxation as 'plant', while at the same time constituting 'new build' and therefore zero-rated for the purposes of VAT. At the same time, the engineer was to include it in certificates he issued to the defendants (Scicon) for payment to the contractors. It was in connection with this that the judge made the remark: 'An engineer who includes in a certificate goods which are not the property of the contractor does so at his own peril.'

In the particular circumstances of this case, no doubt the employer would be estopped from complaining that the engineer, in issuing certificates for goods the title for which he well knew was not intended to pass to his employer, was negligent, and the judge held that the property *had* in fact passed when the air-conditioning became incorporated in the realty.

The architect who included the slates in his interim certificate in *Dawber Williamson* v. *Humberside County Council* (1980) could possibly have a similar defence against his employer — namely, that the employer, by entering into JCT 63, was estopped from claiming that his architect was negligent in certifying sums for goods of which the employer could not become owner.

It is clear however that an architect may be liable if he includes in his interim certificates the value of defective work or unfixed materials for which the contractor has no title.

6.02 Interim certificates: valuation

JCT 80 clause 30.1.3 requires the architect to issue interim certificates to the contractor at the intervals stated in the Appendix. If none is stated the interim certificates are to be issued at intervals of one month.

The amounts which are to be included in those certificates are:

(a) The total value of work properly executed by the contractor up to the relevant date [6.15].

(b) The total value of materials and goods delivered to or adjacent to the works for incorporation therein.

By clause 16.1 when such unfixed materials and goods are included in an interim certificate and that is paid to the contractor 'such materials and goods shall become the property of the employer'.

(c) In addition, the architect has a discretion under clause 30.3 to include in an interim certificate off-site materials and goods, subject to the conditions set out in that clause.

The provision under JCT 80 clause 30.1.3 for the compulsory inclusion of on-site materials and goods raises difficulties. The goods may have been delivered to the contractor subject to a retention of title clause by the supplier. In that event ownership of the goods will only pass to the contractor if he has paid for them, or to the employer when they become permanently affixed to the realty. Or the unfixed materials may belong to a work and materials sub-contractor, in which case it will never be the intention of the parties that the contractor should have a title to pass to the employer under clause 16, so that the situation which arose in *Dawber Williamson* v. *Humberside County Council* can easily recur [2.03].

6.03 Retention of title on the sale of goods

It is not proposed to expand greatly on the question of retention of title since the present author has dealt with the current law fully in his book *Retention of Title on the Sale of Goods* (1982). Retention of title clauses, which are recognised expressly by section 19(1) of the Sale of Goods Act 1979, have been recognised as valid ever since the House of Lords case of *McEntire and Maconchy* v. *Crossley Brothers Ltd* in 1895. The full value of such clauses, however, did not occur to manufacturers and suppliers until the case of *Aluminium Industrie Vassen* v. *Romalpa* in 1976. Since then they have been widely used, not least by manufacturers of building products, who have learnt by bitter experience that if there is any insolvency in connection with a construction contract, it is they who suffer most of all as unsecured creditors, mainly receiving any dividend in the liquidation even though their materials may have been incorporated in the fabric of a building and are of value to the owner.

The form of the retention clause varies. A 'simple' retention of title clause may read: 'The property in all goods is not to pass to the customer until they are paid for in full.' A 'current account' clause may retain property until all indebtedness by the customer in respect of all transactions has been discharged. A 'proceeds of sale' clause may seek to make monies received for the goods the property of the vendor, and an 'aggregation' clause may purport to vest the title of any goods admixed with or adjoined to the vendor's goods in the vendor.

Such clauses will, in most cases, be defeated when the goods are affixed to the realty and cease to exist as chattels.

However, it is possible by contract to provide a right for an unpaid vendor to sever such chattels from the realty.

In *Hobson* v. *Gorringe* (1897) it was held that the owner of goods supplied on hire purchase was entitled to enter on the land and remove his goods which had been affixed to the realty. The terms of the hire purchase agreement constituted a licence, coupled with an interest in the land. Such a clause may be effective against the other contracting party, i.e. the contractor in possession of the site and his liquidator or receiver, but will rarely defeat debenture holders with floating charges or mortgagees of specific property with charges created before the addition of the chattel to the realty: *Longbottom* v. *Berry* (1869); *Meux* v. *Jacob* (1875). If the charges are created after affixation, it would appear that the right to enter and recover the goods, even affixed goods, takes priority over subsequently created equitable interests, whether the debenture holders are aware of the hire purchase agreement or not: *re Morrison Jones & Taylor Ltd* (1914).

It is therefore open to a supplier to write a clause such that, even where goods such as doors have been affixed to the realty, an unpaid vendor would have the right to remove them in the event of the contractor's insolvency.

A retention of title clause is not a charge which requires registration under section 95 of the Companies Act 1948: *Clough Mill* v. *Martin* (1984) (CA).

6.04 Retention of title and the Joint Contracts Tribunal

In 1978 the Joint Contracts Tribunal issued a formal notice entitled 'Retention of title (ownership) by suppliers of building materials and goods'.

> 'The Joint Contracts Tribunal announces that, through its constituent bodies it has been informed of the following problem: Some suppliers of building materials and goods are including provisions in their contracts of sale with contractors and sub-contractors under which the supplier retains ownership of such goods and materials after their delivery to the site: The terms on which such retention of ownership is secured appear to vary; but in many cases the passing of ownership to a contractor or sub-contractor is dependent upon payment in full for the relevant materials and goods. It is understood that suppliers anticipate being able to use such provisions to enable them either to repossess the goods and materials if they have not been paid for in full; or to claim against the proceeds of any re-sale.
>
> Some employers (and their professional advisers) are seeking to obtain proof of ownership by the contractor (or through the contractor, by any relevant sub-contractor) before operating the provisions of Clause 30(2) [of the 1963 JCT form]. Moreover in current tenders some employers are seeking to amend Clause 30(2) by making it a condition of the operation of the valuation provisions in that sub-clause that the contractor provides proof of ownership.
>
> The tribunal has considered this matter to see if there is sufficient substance for the concern being expressed by some employers (and their professional advisers) to justify any change in the existing Standard Form provisions in Clause 30(2) and in Clause 14(1). The tribunal, with the concurrence of its constituent

bodies, does not think that any change is desirable and the main reasons for reaching this decision are set out below:

Reasons for decision by Tribunal not to amend Clause 30(2) and Clause 14(1)

(1) A requirement on the contractor to prove ownership of on-site materials and goods could raise serious legal problems, both for the contractor, any relevant sub-contractors and also for the employer (and his professional advisers). Such a requirement would, therefore, be difficult to meet and so might mean, in practice, that payment for on-site goods and materials would not be operated. Moreover, the obtaining of proof of ownership would add to administration costs as would the checking of such proof by, or on behalf of, the employer. The tribunal concluded that such a requirement would add to the costs of building work by reason of additional administration; and might cause tender prices to rise because contractors and sub-contractors could no longer be certain that materials and goods properly on site would be valued and paid for in interim certificates.

(2) The degree of risk to the employer from not obtaining proof of ownership before paying for on-site goods and materials in interim certificates was not considered sufficiently great to justify the possible additional costs, referred to in (1) above for the following reasons:

(a) The period of risk runs only until such time as the on-site goods and materials are incorporated in the works; from the time of incorporation they cease to be chattels, and any right to re-possess by a supplier would be lost. The period of risk is, therefore, from date of payment by the employer of the relevant interim certificates to the time at which the relevant goods and materials are incorporated in the works. This is unlikely to be more than a relatively short period.

(b) During the limited period referred to in (a), the risk of re-possession by a supplier would only, in practice, arise if a main contractor became insolvent. Such insolvency occurs only on a small proportion of the total number of building contracts and this reduces the degree of risk even further.

(c) The tribunal understands that in many cases the supply contract permits the contractor or sub-contractor to re-sell the goods and materials. In such cases the supplier's rights are

> against the proceeds of re-sale and the supplier has no right to re-possess the goods and materials. This reduces the risk to the employer still further.'

Subsequently, the Joint Contracts Tribunal had second thoughts and issued the First Amendment to the 1980 Form, which has been discussed in [2.03]. However, it becomes clear from the conditions, subject to which an architect can include off-site materials in his interim certificates, that he should be alert to the possibility of liability if he includes on-site materials in his certificate where he knows or ought reasonably to have known that the contractor has no title to pass under clause 16 to the employer.

It could be added that an architect cannot necessarily know whether or not any contractor or supplier has title in any materials. Some vendor further back in a string of sale contracts may have retained title. Before the law was changed it was not infrequent for a whole chain of subsequent innocent purchasers to be brought into court when some villain had sold a car subject to hire purchase, because none had a title and the hire purchase company had seized the car from the last in the chain.

6.05 Interim certificates: off-site materials

JCT 80 empowers, but does not require, the architect to include in an interim certificate the value of any materials or goods 'before delivery thereof to or adjacent to the works' provided they are materials intended for incorporation (clause 30.3).

There are a number of conditions which circumscribe this discretion of the architect:

30.3.4 'where the materials were ordered from a supplier by the Contractor or by any sub-contractor, the contract for their supply is in writing and *expressly provides that the property therein shall pass unconditionally to the Contractor or sub-contractor* . . . [author's italics]

.5 where the materials were ordered from a supplier by any sub-contractor, the relevant sub-contract between the Contractor and the sub-contractor is in writing and expressly provides that on the property in the material passing to the sub-contractor the same shall immediately pass to the Contractor;

.6 where the materials were manufactured or assembled by any sub-contractor, the sub-contract is in writing and expressly provides that the property in the materials shall pass unconditionally to the Contractor . . .'

These clauses are specifically aimed at preventing the architect from including in his interim certificates materials off-site for which the original supplier has retained title until payment.

Curiously, these clauses would appear to exclude the architect from including in interim certificates off-site materials prepared or obtained by a nominated sub-contractor under the 'basic method', NSC/4, or the 'alternative method', NSC/4a, since neither of these contracts provides that property in the goods shall ever pass to the contractor.

A further requisite before the architect may include off-site goods or materials in any interim certificate is:

30.3.8 'the Contractor provides the Architect with reasonable proof that the property in the materials is in him and that the appropriate conditions set out in clauses 30.3.1 to .7 have been complied with.'

The sub-contractor enters into a similar obligation to the contractor under NSC/4 clause 21.2.3. This being so, it seems strange that the provision in clause 30.3 appears at all in the contract since there can rarely be cases where the condition can be complied with. The clause corresponds roughly to clause 30(2A), incorporated at a late stage into the JCT 63 contract.

But it would be effective evidence against an architect of the danger of including in his interim certificate on-site materials without proof that the contractor owns them — but no doubt the architect could argue as against his employer that, because the employer has signed JCT 80 which says in clause 30 that the architect *shall* include such materials in his interim certificate, the employer is estopped from alleging that the architect has been negligent by so doing. This will almost certainly be so where the employer is a local authority, but a private employer who is using JCT 80 on his architect's advice would not be estopped.

6.06 Interim certificates: sub-contractors

JCT 80 clause 30.2.1 also requires the architect to include in his interim certificates the total value of sub-contract works properly executed by sub-contractors and the total value of the materials and goods delivered to or adjacent to the Works by sub-contractors.

He may also include, subject to similar conditions as in clause 30.3, off-site materials and goods for incorporation by sub-contractors.

He is obliged to include the contractor's profit element in respect of payments to sub-contractors.

JCT 80 clause 35.13.1 requires the architect, on the issue of each interim certificate, to advise the contractor and each nominated sub-contractor of the amount contained in the certificate in respect of the work of each sub-contractor. He is apparently not called upon to advise domestic sub-contractors of the amounts due to them contained in the certificate.

Before the issue of all certificates other than the first, the contractor is required to produce to the architect evidence that he has paid these sums to the nominated sub-contractors (again, apparently, not to domestic sub-contractors).

If the contractor fails to provide reasonable proof of such payment, 'the architect shall issue a certificate to that effect' with a copy to the nominated sub-contractor concerned: clause 35.13.5.2. Thereafter the employer may deduct that sum from future payments to the contractor and when he has done so, must pay the same direct to the nominated sub-contractor concerned.

But the employer is under no obligation to pay to nominated sub-contractors more than the deduction he is able to make from the next interim certificate to the contractor: JCT 80 clause 35.13.5.3.

Moreover, the whole machinery should clearly not be operated if the reason why the contractor fails to provide proof of payment is because it is in fact due to the default of the sub-contractor.

This procedure is permissive only in the case of all nominated sub-contractors but is mandatory in the case of those who have entered into NSC/2 or 2a with the employer. Provision is made in JCT 80 clause 35.13.5.4.2 regarding retention monies and in clause 35.13.5.4.3 regarding pro rata distribution where two or more nominated sub-contractors have to be paid direct.

But this right to pay a sub-contractor direct ceases in the circumstances where it would be most valuable to nominated sub-contractors, namely where the contracting company has had a petition presented for its winding up or has passed a resolution for winding up other than for the purposes of amalgamation or reconstruction: JCT 80 clause 35.13.5.4.4.

In the case of sub-contractors who enter into NSC/2 (but not NSC/2a) there is an obligation to repay to the employer monies paid direct to the sub-contractor if the employer produces reasonable proof that at the time when the payment was made there was in existence a petition or a resolution for the winding-up of the contractor: NSC/2 clause 7.2.

Unfortunately all these provisions stem from an erroneous view of the law. In the *JCT Guide* at page 24 the explanation for this is given as follows:

> 'The reason for this change from the 1963 Edition is that the 1963 provision was open to challenge by the . . . liquidator of the contractor and it was therefore considered inappropriate to include it in the 1980 edition.'

This appears to be based on the assumption that *British Eagle* v. *Air France* (1975) has somehow overruled *re Tout and Finch* (1954), which dealt expressly with clause 21(c) of the then RIBA contract and provided that, in default of the contractor producing evidence that he has paid a nominated sub-contractor, the employer could pay the nominated sub-contractor directly and 'deduct the amount so paid from any sums otherwise payable to the contractor'. It was held by Mr Justice Wynn-Parry that clause 21(c) applied to both interim certificates and final certificates, 'that notwithstanding the liquidation of the contractor' it was binding on the liquidator and that the employer concerned was entitled to pay the sub-contractor the balance remaining unpaid of certified sums — as had been previously held regarding similar provisions in the ICE contract in *re Wilkinson ex parte Fowler* (1905).

The *British Eagle* case dealt with contractual arrangements which might be taken to alter the order in which assets of insolvent companies are distributed under sections 302, 312 and 319 of the Companies Act 1948. At issue was a pooled fund of air fares which were to be distributed by periodical clearings. Nothing could be further removed from the provisions existing for so many years previously under RIBA and JCT contracts. There is all the difference in the world between a common fund in which receipts are mingled and a situation where the contractor is nothing more than a conduit pipe in the machinery of transferring from an employer to a nominated sub-contractor monies to which he is entitled from the moment of certification.

6.07 Interim certificates: work 'properly executed'

In *Sutcliffe* v. *Chippendale and Edmondson* (1971), Judge Stabb, sitting on Official Referee's business, said in relation to the issue of interim certificates, after quoting JCT 63 clause 30:

> 'I have had the advantage of . . . listening to the views of no less than five experienced independent architects and quantity surveyors . . . in addition to the two architects and two quantity surveyors involved in the case.
>
> It is clear that there is a variation within the profession as to the practice in the preparation of the certificate, which is only to be expected, and also some divergence of principle as to what the amount certified is designed to represent, which is perhaps more surprising.'

One architect told his lordship that it was his practice, as a matter of routine, when he paid a site visit if he observed defective work to issue a written order to the contractor, with a copy to the quantity surveyor so that he would know not to include it in his next certificate. Without such a system, he said, the architect would have to get in touch with the quantity surveyor in some other way to pass on the information.

The judge said of this witness:

> 'He took the view that the certificate should conform as strictly as possible to the terms of the contract and that, if he was not satisfied with any particular item of work, it was implicit that the value of the work was not to be included in a certificate covering work properly executed.'

Other witnesses, the judge said, expressed the view:

> 'that such a certificate was more of an approximation of the value of the work as it progressed, assessed by the quantity surveyor without any detailed inspection of the work, the object being

simply to provide a reasonable progress payment for the contractor based upon a comparatively cursory examination of the site.

All were agreed that responsibility for the detection and, if necessary, the exclusion from the certificate of defective work was that of the architect as opposed to the quantity surveyor, whose concern was as to quantity and not quality.

[The judge concluded:] Faced with these two opposing views held by men of experience in the profession, my task of deciding what is an architect's proper professional function in this respect is plainly a difficult one.

Since everyone agreed that the quality of the work was always the responsibility of the architect and never that of the quantity surveyor and since work properly executed is the work for which a progress payment is being recommended, I think that the architect is in duty bound to notify the quantity surveyor in advance of any work which he, the architect, classifies as not properly executed, so as to give the quantity surveyor the opportunity of excluding it.

As to the system or method of communication between architect and quantity surveyor to be adopted, I make no comment, save to say that for a busy architect merely to rely upon his memory for this purpose seems, in my view, to be unsatisfactory.

Furthermore, I can well understand an architect's reluctance to devote too much time to matters concerned with certificates at interim stage, since, in the normal course of events, work that may be defective and unacceptable at that stage, can and will be remedied at a later stage, and therefore over-certification and consequent overpayment by his employer would only be temporary and must automatically adjust itself as the contract reaches completion, and defects are remedied by the contractor at his own cost before any final certificate is issued.

But should circumstances unexpectedly bring the contract to a premature end whilst such overpayment remains uncorrected, it is difficult to see how the architect can avoid responsibility, if the overpayment proves to be irrecoverable from the contractor as has happened in this case.

In my considered opinion the strict approach is the right one.

If meaning is to be given to the words 'work properly executed' in clause 30 sub-clause 2, I cannot see how the architect can avoid the requirement to exclude work, which is not properly executed, from the value of the work for which he recommends his employer to make payment on account, and if work is defective and unacceptable as it then stands, it must, in my view, be classified as work not properly executed until the defect has been remedied.

I do not accept that the words "work properly executed" can include work not then properly executed but which it is expected, however confidently, that the contractor will remedy in due course.

So long as the contractual basis of the certificate is the valuation of the work *properly* executed, the architect should first satisfy himself as to the acceptable quality of the work before requiring his employer by way of certificate of make payment for it, and in particular should keep the quantity surveyor continually informed of any defective or improperly executed work which he has observed.'

JCT 80 clause 30.10 specifically provides that the contractor cannot rely on any certificate of the architect, except the final certificate in some circumstances [6.19], as conclusive evidence that any works or materials to which it relates are in accordance with the contract.

6.08 Interim certificates: the architect's role

In *Sutcliffe* v. *Chippendale and Edmondson* (1971) Judge Stabb discussed whether the quasi-arbitrator privilege which then protected an architect when giving a final certificate applied equally to interim certificates. He quoted Mr Justice Sankey in *Wisbech Rural District Council* v. *Ward* (1927).

'Although it is probably right to say that in giving a final certificate the architect acts in a quasi-judicial character unless there is some express clause in the contract to contradict it, it cannot, I think, be asserted that in giving an interim certificate he is so acting. Personally, I should have thought that the inference was just the other way — namely, that in giving an interim certificate he is merely acting as an agent for the building owner unless there is something in the contract to contradict that.'

Judge Stabb said:

'Plainly, it is part of an architect's duty of supervision, as agent for his employer, to see that the work is properly executed, and therefore to my mind supervision and the issuing of interim certificates cannot be regarded as wholly separate and distinct functions.

> I think that it was rightly contended on behalf of the plaintiff that in a well supervised contract an architect would not certify for work not properly executed.
>
> I have come to the conclusion that the architect, in discharging his function of issuing interim certificates, is primarily acting in the protection of his employer's interests, by determining what payment he can properly make on account, such determination being based upon his assessment of the value of the work properly carried out, that assessment being perofmred by virtue of his professional skill, for which the building owner has engaged him.
>
> It is in my view part of his supervisory function to see that the value of work properly executed, and only work properly executed, is included in the valuation for the purpose of an interim certificate, and that therefore he is under a duty of care to his employer in the performance of that function.
>
> Accordingly, if by his failure to exercise due skill and care, he should fail to exclude from the certificate the value of work not properly executed, he would in my view be liable to his employer for any damage attributable to such default.'

He also took the view that the architects in this particular case:

> 'must have known at least of the possibility that these contractors would not be recalled to remedy those defects. They knew that the contract had been nearing completion.
>
> In these circumstances, I consider that it was their plain duty to be particularly accurate in their valuation of the work properly executed, to that date.

This is the same case as that which subsequently went to the House of Lords under the name of *Sutcliffe* v. *Thackrah* (1974), and in which it was held that an architect giving interim certificates was not protected from actions for negligence by reason of being a 'quasi-arbitrator'.

6.09 Retention monies: the amount

All the sums so far referred to as being payable in an interim certificate are subject to deduction of retention monies at the agreed rate. The purpose of this is to provide a fund from which the employer is entitled to have defects rectified.

The amount specified in JCT 80 clause 30.4.1.1 is to be 5 per cent unless a lower rate has been agreed and specified in the Appendix under the heading 'Retention Percentage (if less than 5 per cent)'.

A footnote suggests that 'where the Employer at tender stage estimates the contract sum to be £500,000 or over, the Retention Percentage should not be more than 3 per cent'. This, of course, creates no contractual obligation in the absence of a separate and specific agreement on this point incorporated in the appropriate section, mentioned above, of the Appendix.

For work performed or otherwise chargeable on an interim certificate between practical completion [11.01] and the issue of a certificate of completion of making good defects [11.05], the retention is to be at half the agreed rate.

6.10 Retention monies: a trust fund

Clause 30.5 provides that the employer's interest in the retention monies 'is fiduciary as trust for the Contractor and for any nominated sub-contractor'. At the date of each interim certificate, the architect is to issue a statement specifying all retention monies retained for the contractor and each sub-contractor and they are all to be supplied with a copy.

The essence of JCT 80 clause 30.5.1 is that retention money is the contractor's money. The employer does not own it. He has no legal or equitable interest in it. He merely holds it as trustee for the contractor. Important implications of law follow from this.

The intention, of course, is to create a situation such that if the employer becomes insolvent, retention monies in his hands are not his property: they are branded with the proprietary interests of the contractor or sub-contractor beneficiary, and a receiver or liquidator or trustee in bankruptcy cannot lay his hands on them.

But once a trust is created certain effects inevitably follow.

Firstly, the trustee is obliged to invest trust monies in such investments as are prescribed by the Trustee Investments Act 1961. The words of JCT 80 clause 30.5.1 'but without obligation to invest' purport to exonerate the employer from the obligations of that Act and of the Trustee Act 1925.

6.11 Separate fund necessary for retention monies

Furthermore a trustee is under an obligation not to mingle trust monies with his own. This means that he *must* establish a separate trustee account. In *Rayack Construction Ltd* v. *Lampeter Meat Co. Ltd* (1979) the plaintiffs entered into a contract in 1978 in the JCT 63 form which contained clause 30(4) in the same terms as JCT 80 clause 30.5.1. They sought a declaration that the defendants, their employers, were obliged to pay the retained monies into a separate bank account to be applied only in accordance with the trust specified in clause 30(4).

The employers claimed that JCT 63 clause 30(4) contained no express provision requiring them to establish a separate fund.

Mr Justice Vinelott in the Chancery Division accepted that clause 30(4) required the employers to establish a separate fund.

In the action Rayack Construction Ltd claimed a declaration that: '[Lampeter Meat Co. Ltd] are obliged forthwith to pay the sum of £336,500, together with all future retention monies as and when they are certified . . . into a separate bank account thereto to be applied only in accordance with the trust specified in clause 30(4)' of the JCT 63 contract. They further sought a mandatory injunction requiring that this should be done forthwith, and an injunction restraining the defendants from applying future retention monies in any other fashion.

On behalf of the defendants, reliance was placed upon the opinion expressed by I. N. Duncan Wallace, editor of *Hudson's Building and Engineering Contracts,* 10th Edition at page 729 where he said:

> 'It does not however appear to have been appreciated that retention monies for the main contractor are never likely to be an identifiable fund in the hands of an employer to which the trust could attach. In the absence of such a fund the provision amounts to little more than an attempt to obtain a preferential debt for the main contractor.'

Mr Justice Vinelott did not accept Mr Duncan Wallace's version of what clause 30(4) meant. Indeed it is difficult to see how, because a trustee has not done what equity requires him to do, that can be an argument for the contention that there is no trust. But the sentence quoted was taken by counsel out of context, for the real burden of the argument in *Hudson*, 10th Edition, page 730, is whether sub-contractor's retention can be effectively traced if it has been paid into an insolvent contractor's overdrawn bank account, when the bank will have prior claims.

Mr Justice Vinelott said:

> 'In my judgment, clause 30(4) construed in the context of the agreement as a whole, does impose an obligation on an employer to appropriate and set aside as a separate trust fund retention monies.
>
> Unless clause 30(4) is construed as imposing such an obligation, it cannot have any practical operation.
>
> Further, clause 30(4) refers to the "contractor's beneficial interest therein" and the predicated beneficial interest could only subsist in a fund so appropriated and set aside.
>
> It is, in my judgment, clear that the purpose of retention under clause 30(4) is to protect both employer and contractor against the insolvency of either.
>
> [Counsel for the defendants] said that they might be faced with a cash flow problem if they were now compelled to appropriate and set aside these monies. That argument is double-edged.
>
> It would be wrong to expose the plaintiffs to any degree of jeopardy in order that the defendants might continue to use in their business monies which ought to have been appropriated and set aside for the security of the plaintiffs.'

In other words, in all circumstances, contractors are entitled to have separate accounts set up in respect of retention monies as trust funds.

If this is not done, equity will treat as done that which ought to be done. In *re Arthur Sanders Ltd* (1981) Mr Justice Nourse in a case on JCT 63 and the NFBTE/FASS Green Form held that:

> 'Once the sums notionally set aside have been impressed with the relevant trusts, they remain subject to these trusts, whatever the fate of the contractor's employment or of the contract itself.'

Similarly, it followed that where JCT 63 was entered into between the contractor and the employer and the NFBTE/FASS form between the contractor and the sub-contractor, 'it creates a trust in favour of the sub-contractor'.

The present editors of *Building Law Reports* suggest in a note to the *Rayack Construction* case *supra* that since all injunctions, mandatory or prohibitive, are discretionary, another court might in different circumstances not order separate funds for retention monies. But no court of equity could by exercise of discretion exonerate any trustee from the fundamental obligations of being a trustee, even with the consent of the beneficiaries.

Although all beneficiaries can in some circumstances alter the terms of a trust by deed, they can never alter the fundamental obligations of trustees established over the years by the courts of equity.

6.12 Contractor entitled to interest on retention

The third obligation on trustees is never themselves to benefit from trust monies. The trustee is not entitled to remuneration from a trust. Only a solicitor trustee with a charging clause in the trust may under the rule in *Cradock* v. *Piper* (1850), if he is acting for himself and other trustees, charge the costs of so acting; and then only if his costs would have been no greater than if he were acting for the lay trustees alone.

It follows from this that no trustee is ever entitled to profit from his position of trust. In one case, where the trustees of an estate went shooting grouse on it, they were held liable to account to the beneficiaries for the value of the shooting. A stockbroker's clerk who was trustee of an estate was not entitled to commission from his employers when they were asked to value it. The test is not whether the trust has suffered loss by the trustees' action. It is whether an individual has benefited by his position of trustee.

In *Swain* v. *Law Society* (1981) the Law Society, being in a fiduciary position, was held liable to repay to individual members commission it had received from the insurance brokers on its members' compulsory insurance. There is no escape from this rule that trustees or those in fiduciary positions analogous to trustees must not benefit from the trust.

As Lord Justice Stephenson said in the Court of Appeal in the *Swain* case:

> 'The Law Society obtained the insurance contract by reason of its fiduciary position. . . . The commission was acquired by the Society using its fiduciary position. It is therefore accountable for it to the solicitor beneficiaries.
>
> There is *an inflexible rule of equity*, exemplified in *Phipps* v. *Boardman* (1967) . . . that a person who owes a fiduciary duty to others is accountable to them for any profit or other advantage he obtains from his fiduciary position' [author's italics].

Although the House of Lords subsequently set aside the judgment, it did not dispute the description of the obligations of trustees but did so on the ground that the Law Society, in setting up the insurance scheme, were exercising statutory powers and were not in the position of ordinary trustees.

There is an exact analogy between that and employers in possession of contractors' retention monies.

'An indelible incident of trust property is that a trustee can never make use of it for his own benefit,' said Lord Cottenham LC in *Foley* v. *Hill* as long ago as 1848. He cannot do so even if the beneficiary consents. There can be no doubt that under JCT 63, in view of the silence of the contract, the contractor was entitled to all interest earned on retention monies.

JCT 80 clause 30.5.3 in the local authorities edition reads: 'Number not used'. In the private edition clause 30.5.3 empowers the contractor to call on the employer to establish a separate fund for retention monies held but purports to authorise the employer to receive interest on that fund.

Although common lawyers rarely appreciate the fact that one cannot by contract purport to interfere with trust principles, chancery lawyers do. It must follow that JCT 80 clause 30.5.9 cannot confer the right on the employer to receive interest on retention monies which are trust funds. There is at present no direct authority to support this proposition but it is submitted that it is a necessary corollary of the two cases already cited.

Not all retention money is trust money. The JCT Agreement for Minor Building Works for example, does not expressly create a trust in favour of the contractor, and in these circumstances the employer will be entitled to interest on the money. In the event of the employer under the JCT Minor Works Form becoming insolvent, the retention money would rank solely as a debt, along with those of other unsecured creditors, due from the employer to the contractor.

Under the JCT Intermediate Form the employer is trustee of retention monies for the employer, but under the sub-contract forms the contractor is not trustee of retention monies for the sub-contractor — a strange situation.

6.13 Interim certificates: other sums

In addition to the amounts already specified above, all of which are subject to the deduction of retention monies, the contract provides for other sums to be included which are not subject to such deductions.

JCT 80 clause 3, a new clause, provides that where any amount is to be added or deducted from the contract sum, or dealt with by an adjustment of it, these amounts are to be included in the next interim certificate after the time when 'such amount is ascertained in whole or in part'.

The following appear to be covered by this:

— fees or charges, rates and taxes (other than VAT): JCT 80 clause 6.2
— architect's instructions concerning setting out: JCT 80 clause 7
— cost of unnecessary opening up: JCT 80 clause 8.3
— payments in respect of patents: JCT 80 clause 9.2

and where the architect allows:

— contractor not to rectify faults: JCT 80 clause 17.2 and clause 17.3; but see [6.12]
— sum expended by contractor under JCT 80 clause 21.2 insurance: JCT 80 clause 21.2.3.

In addition the following are to be included under similar conditions in the interim certificate:

— direct loss and/or expense caused by matters materially affecting the regular progress of the work: JCT 80 clause 26.1
— direct loss and/or expense caused by discovery of antiquities: JCT 80 clause 34
— fluctuations payable under JCT 80 clause 38 or 39, whichever is applicable.

It will be noted that this particular clause makes no provision for sums due under JCT 80 clause 40, the Price Adjustment Formula, to be added to the interim certificate. This is covered by JCT 80 clause 30.2.1.1 which makes such payment subject to deduction of retention money.

The matter, erratically, is also dealt with in the Formula clause:

40.1.1.1 'The Contract Sum shall be adjusted in accordance with the provisions of clause 40. . . .'

40.1.3 'The adjustment referred to in clause 40 shall be effected . . . in all certificates for payment issued under the provisions of the Conditions.'

40.2 'Interim valuations shall be made before the issue of each Interim Certificate.'

Finally the interim certificate must include sums due to those nominated sub-contractors, who have entered into NSC/2 or NSC/2a, where final payment is to be made in accordance with the provisions under JCT 80 clause 35.17, and sums due to sub-contractors under NSC/4 or NSC/4a for a multitude of matters set out in NSC/4 or 4a clause 21.4.2, itself a repertory of repetends.

6.14 Interim certificates: deductions by architect

The architect is under a duty to deduct from the sum named in his interim certificate monies due to the employer which are not subject to retention. These specifically include:

— insurance premium which the contractor has failed to pay: JCT 80 clause 22A.2
— liability of contractor for substituted sub-contractor: JCT 80 clause 35.18.1.2
— liability of contractor to employer where sub-contractor has withdrawn as result of contractor's default: JCT 80 clause 35.24.6
— 'any amount allowable to the Employer under Fluctuation Clauses': JCT 80 clause 38 or 39.

The architect is not entitled to pay for work which is not in accordance with the contract and, it is submitted under the principles of *Basten* v. *Butter* (1806), he can abate the sum awarded in respect of sub-standard work which is of less value than if it complied with the contract. As was said in a case which illustrates the distinction between abatement of value and damages for breach, *Mondel* v. *Steel* (1841):

> 'The defendant is now permitted to show that the chattel, by reason of non-compliance with the warranty in the one case, and the work in consequence of the non-performance of the contract in the other, were diminished in value: *Kist* v. *Atkinson* (1809); *Thornton* v. *Place* (1832).'

It is clear that interim certificates are continuing valuations, and if it is discovered that defective work has been included and paid for in an earlier certificate, the architect is under a duty to deduct it in a subsequent one.

Moreover clause 7, which deals with setting out the works, allows the architect to provide that the contractor shall not amend any of his errors 'in which case the contract sum shall be adjusted accordingly'. Far from providing additional sums for the contractor, as JCT 80 clause 30.2.2.1 would seem to suggest, this must mean that if the architect allows the contractor to get away without correcting his errors in setting out, the contract sum should be corrected by deducting the costs of correction which the contractor would otherwise have to bear, or at least the lesser value of the work.

It is also submitted that there is power to make deductions where, under JCT 80 clauses 17.2 and 17.3, the architect has similarly dispensed with the contractor correcting defects 'in which case the contract sum shall be adjusted accordingly'.

To read this clause in the same way as JCT 80 clause 30.2.2 and (for final certificate) clause 30.6.2.10 appear to do as representing only *additions* to the contract sum in favour of the contractor is surely not right. Clearly, what is intended here is in JCT 80 clause 7: that this contract sum shall be *reduced* by the costs that the contractor would otherwise have to pay. To read these clauses as meaning solely that the employer has to pay the cost of remedying the contractor's breach of contract is to abuse language.

Mr Duncan Wallace has said:

> 'There is no power to accept unremedied work accompanied by an abatement price, but only in some unspecified circumstances to make the employer pay for breach of contract by the main contractor or nominated sub-contractor.'

It is submitted that this is not the view the courts would take of the words 'in which case the contract sum shall be adjusted accordingly'. And if it was the view held by the draftsmen of JCT 80 and expressed in clauses 30.2.2.1 and 30.6.2.11, it is submitted that they are in error as to the meaning and intention of their own words in JCT 80 clauses 7 and 17.

As a matter of principle and of valuation, the architect is entitled to reduce the sum due to the contractor for his default and breach of contract in setting out or in creating defects, by work which is not in accordance with the contract, without requiring the contractor to rectify.

JCT 80 clause 30.2 cannot therefore be regarded as a comprehensive and exhaustive review of all the sums that an architect may be entitled to deduct from the amount he would otherwise have certified, since the contract elsewhere expressly provides otherwise.

6.15 Changes in interim certificates procedure: JCT 80 clause 30

A contractor is now entitled to be paid within 14 days from *the issue* of each interim certificate. The contract does not specify what is meant by that, but it would appear not to be the date which the architect places on it, but the actual date on which he despatches it.

Formerly, under the private edition he had to send an interim certificate to the contractor, who had to present it to the employer, who then had 14 days from the day on which he received it to pay it. Now under JCT 80 clause 5.8 all certificates, unless separately specified, have to be issued to the employer, with a copy to the contractor.

The total amount stated as being due on any one certificate is now to be less the total amount in interim certificates previously issued. Formerly, under JCT 63 it was less any amounts previously *paid*, which imposed an obligation on the architect to inquire whether his earlier certificates had been paid or not and to include them if they had not in his current certificate. In spite of that express provision in JCT 63 architects largely ignored it, and the RIBA Interim Certificate did not correspond with the obligation of JCT 63 in that it provided for the architect's valuation to be subject to the deduction of the sums previously *certified*.

As before, the account is to be made up to a date not more than seven days before *the date* of the interim certificate: JCT 80 clause 30.2.

A distinction therefore appears to be drawn in JCT 80 clause 30 between 'the date of issue of the certificate' (clause 30.1.1.1) and the date of the certificate (clause 30.2).

JCT 80 clause 30.1.2 now provides that 'interim valuations shall be made *by the Quantity Surveyor* whenever the Architect considers them to be necessary'. How an architect can certify 'the total value of the work properly executed' (as required by JCT 80 clause 30.2.1.1) without a valuation remains a mystery. However, the addition of the words 'by the Quantity Surveyor' to the wording of this clause appears to have given rise to a belief that quantity surveyors are entitled to make an extra charge for interim valuations.

Where the Formula Rules for fluctuations are adopted, clause 40.2 provides that the words 'whenever the Architect considers them necessary' are to be deleted from JCT 80 clause 30.1.2. In other words, a valuation by the quantity surveyor always has to precede an interim certificate where the Formula Rules have been adopted.

JCT 80 clause 30.1.3 now requires interim certificates to continue to be issued at the specified intervals 'up to and including the end of the period during which the certificate of Practical Completion is issued.'

Thereafter they shall be issued 'as and when further amounts are ascertainable as payable to the contractor' up to the certificate of making good defects, but not at lesser periods than one month. These represent changes from JCT 63 clause 30(1) to meet contractors' complaints that they often did not get interim certificates once the certificate of practical completion had been issued.

6.16 Employer's right to deduct from certified sums

The employer is entitled by the contract to deduct from sums certified by the architect certain specific and liquidated sums. The most important of these are liquidated damages in the circumstances set out in JCT 80 clause 24.2.

Other sums for which express authority is contained in the contract relate to the following:

- — other persons employed where contractor fails to comply with architect's instructions: JCT 80 clause 4.1
- — insurance premiums paid because of contractor's default: JCT 80 clause 21.1
- — insurance premiums: JCT 80 clause 22A.2
- — on determination of contractor's employment: JCT 80 clause 27.A
- — the Finance Act (No.2) 1975 provisions: JCT 80 clause 21
- — JCT 80 Supplementary VAT agreement
- — direct payments to sub-contractors: JCT 80 clause 35.13.5.3.

These provisions leave open the question of whether the employer has a right to refuse to pay certified sums on the ground that he has an equitable right of set-off in respect of a claim to abate the sum due, or counterclaim for unliquidated or quantified damages. The JCT 80 contract does not expressly deal with such claims, as NSC/4 clause 23 now does.

There are two views about this: one is that the equitable right to set-off is excluded only if there is an express term to that effect. It is sometimes said the the House of Lords' decision in *Gilbert-Ash (Northern) Ltd* v. *Modern Engineering (Bristol) Ltd* (1973) supports that view, but that argument cannot be sustained since in that case there was an express right 'to deduct from any payment certified as due to the sub-contractor . . . claims which . . . the contractor . . . may have against the sub-contractor in connection with this or any other contract'.

Moreover, the subsequent House of Lords' decision in *Mottram Consultants Ltd* v. *Bernard Sunley & Sons Ltd* (1974) in which a majority of three to two of the law lords held that because the contract (similar to JCT 63 but not actually on that form) expressly allowed two deductions from certified sums, none other could be allowed and that by implication the equitable right of set-off was excluded. Moreover, the *Gilbert-Ash* case did no more than establish that each case should be construed according to the terms of the contract in question without any *parti pris* for or against the existence of a right of set-off.

Lord Denning MR has in no way resiled from the position he adopted in *Dawnays* v. *Minter* (1971) where he said:

> 'An interim certificate is to be regarded virtually as cash, like a bill of exchange. It must be honoured. Payment must not be withheld on account of cross-claims, whether good or bad — except so far as the contract specifically provides.'

He repeated this view in *Kilby and Gayford Ltd* v. *Selincourt Ltd* (1973) after being referred to the *Gilbert-Ash* case *supra* and said:

> 'So long as a certificate is good on the face of it and is within the authority given by the contract, then it is in accordance with the conditions. It must be honoured.'

And in subsequent cases up to 1982 he repeated his view that there can be no counter-claims which entitle a contractor to make any deductions from certified sums other than those expressly authorised by the contract [6.17].

As Lord Justice Roskill pointed out in *Frederick Mark* v. *Schield* (1971):

> 'The Master of the Rolls was referring to bills of exchange as a convenient analogy . . . [he] was saying that the debt was of a class which, by reason *of the contractual provisions of the contract* [author's italics], ought not to be allowed to be made the subject of a set-off or counterclaim as a reason for not paying the sum the architect has duly certified . . . I entirely agree with every word of the Master of Rolls' judgment.'

With respect, it is submitted that this is correct. By implication JCT 80, as did its predecessor, excludes any deductions from certified sums other than those specified in the contract. Unfortunately, it is difficult to convince the Masters of the Queen's Bench of this. They, like the law lords, have been over-impressed by an article written by Mr Duncan Wallace in the *Law Quarterly Review* (at 89 LQR 36). As to this, see *Building Law Reports*, Vol 3, page v.

There are, of course, various observations by law lords to the effect that *Dawnays* v. *Minter* has been overruled by the *Gilbert-Ash* case, but Lord Denning MR, right up to his retirement in 1982, did not accept that and, with respect, it is submitted that he was right. As he said in *John Thompson Horseley Bridge Ltd* v. *Wellingborough Steel* (1972):

> 'Once work is certified and the amount is due, it must be paid. It is not to be held up for cross-claims.'

It is greatly to be regretted that the JCT did not resolve the matter once and for all by providing in JCT 80, as they did in NSC/4 clause 23.4, that 'the rights of the parties . . . in respect of set-off are fully set out in the contract . . . and no other rights whatsoever shall be implied as terms of the . . . contract relating to set-off'. Even that is clumsily worded and does not deal with the real point. Contractors should seek to have an express term inserted in the contract that:

> 'Nothing shall be deducted from sums certified as due to the contractor by the architect save as expressly authorised herein and payment of such sums shall not be withheld on account of any other claims of any nature whatsoever made by the employer against the contractor.'

Although it would be satisfactory for the conflicting opinions to be resolved by a decision of the Court of Appeal, it would be better if the JCT were to put the matter beyond dispute.

Since the opinion expressed by the present author in the first edition of this book was misrepresented by one reviewer when the book was first published, it should perhaps be stated succinctly: whether an equitable set-off for a counterclaim can be allowed against certified sums is a matter of the construction of the contract; contracts can exclude equitable set-off by necessary implication — as in the *Mottram* case — as well as expressly; and where certain specified deductions are expressly permitted, the implication is that none other is: *expressio unius personae vel rei, est exclusio alterius*. 'The express mention of one thing is the exclusion of another.'

However, the matter merits further discussion.

Common law and equitable set-off

It is proposed to deal further with this vexed question as to whether an employer can, under JCT 80, deduct from certified sums monies which he claims are due to him on counterclaims. The matter is also discussed in Parris: *Arbitration: Principles and Practice* (1983), pages 10–14, in some detail in connection with what is, or is not, a dispute which can be referred to arbitration.

In this, as in all other questions of law, the history of the matter is of the utmost importance. There never was a 'common law' right of set-off, for the simple reason that until the start of the nineteenth century contractual promises were enforceable only through the action of *assumpsit*, which meant there had to be a separate action on the promises of each party. This situation was modified by the Statute of Set-off 1728 which was intended to last for only five years. It became known informally as the 'Insolvent Debtors Act' because it created a right of set-off in those circumstances. This is now dealt with only under the Bankruptcy and Insolvency Acts.

There was another Statute of Set-off in 1734 which lasted until 1879. It is unnecessary to go into details of all its provisions but it applied only where the plaintiff and defendant were mutually *indebted*, i.e. each had a claim for a liquidated sum immediately due: *Crawford* v. *Stirling* (1802); *Morley* v. *Inglis* (1837). Those debts had to exist when the plaintiff sued, when the defence was filed and when the trial took place: *Evans* v. *Prosser* (1789). Under the statute, therefore, there could be no set-off of a claim for liquidated damages, still less for unliquidated damages, which was the nature of the counterclaim in *Dawnays* v. *Minter* (1971).

By the start of the nineteenth century, the common law would allow the purchaser of goods an abatement of price for breach of warranty as to quality: *Kist* v. *Atkinson* (1809); *Thornton* v. *Place* (1832). But again, there was a firm rule that a counterclaim for damages of any kind could not be set-off: *Mondel* v. *Steel* (1841).

Equitable set-off was such as was allowed by the Court of Chancery before its abolition under the Judicature Act of 1873. It could be a counterclaim for unliquidated damages, but whatever its nature it was only allowed in circumstances where the Court of Chancery would have granted an injunction to restrain a plaintiff from enforcing a judgment he had obtained at common law against the defendant. That is, the nature of the defendant's counterclaim had to vitiate entirely the judgment that the plaintiff had obtained or could obtain from the common law courts. The usual example was a debt incurred by the defendant to the plaintiff as a result of fraud by the plaintiff. Examples in the seventeenth and eighteenth centuries were the so-called 'commodity loans', a money-lending device used to entrap sons of wealthy fathers into debt, by purporting to deliver commodities to them which never in fact came into their possession but were immediately sold off at a knock-down price by the money lender.

Equitable set-off must 'directly impeach the right to payment of the debt', it was said.

The Judicature Act 1873 preserved this right to equitable set-off by section 24 but it passed into complete disuse for nearly 70 years, until it was resurrected in 1948 by the case of *Morgan & Sons* v. *S Martin Johnson & Co*.

As Master Bickford-Smith commented in a lecture on the subject:

> 'When resurrected, the doctrine gave rise to considerable difficulties, both theoretical and practical . . . nobody could remember what the old Court of Chancery did.'

Equitable set-off disappeared again for nearly ten years, until it was once again revived in the case of *Hanak* v. *Green* (1958), which was actually about the costs of building litigation. No reference was made to what had been the practice of the old Court of Chancery, and 'equitable set-off' passed into the mythology of English law. The totally irrational belief of many lawyers is that if an employer (or a contractor) puts forward a counterclaim for damages, however spurious it may be, against a certified sum, he is entitled not to have summary judgment given against him under Order 14.

It is sometimes forgotten that in *Dawnays* v. *Minter*, application was made to the House of Lords for leave to appeal. Although it was on the standard contract, the Green Form, leave was refused. Leading counsel for the applicants conceded that only such sums as the Green Form specifically authorised could be deducted from certified sums. The attitude of the House of Lords was that the decision in *Dawnays* v. *Minter* was plainly right.

Master Bickford-Smith in his lecture said:

> 'It is undoubtedly the commercial intention of the parties that the building owner should pay the contractor as progress payments the amounts certified on interim certificates by the architect, with no deductions save those expressly authorised by the contract.'

He said that he took the view that *Dawnays'* case was never overruled and was right. But he added:

> 'But it is no good relying on that before a Queen's Bench Master. It is too much of a risk from the Master's point of view.'

It is submitted the House of Lords held in *Mottram Consultants Ltd* v. *Bernard Sunley & Sons Ltd* (1974) that only sums which the contract specifically authorised could be deducted from certified sums.

The case of *Gilbert-Ash (Northern) Ltd* v. *Modern Engineering* (1973) in no way overruled *Dawnays* v. *Minter* (1971) in spite of frequent observations to this effect. The opinion of Lord Dilhorne should be analysed in detail. He relied upon a passage in *Halsbury's Laws of England* which said:

> 'the contractor is entitled to immediate payment thereof subject to the terms of the contract and any right of the employer to any set-off or counterclaim (for example for liquidated damages).'

There can be no disputing that all construction contracts for more than a century have provided a right to deduct liquidated damages from certified sums, and had no relevance to the question of equitable set-off from certified sums.

In the event, in the *Gilbert-Ash* case there was express contractual provision for the contractor to deduct any sums claimed on that contract or any other.

6.17 Sums due under an architect's certificate

Even though JCT 80 does not have an arbitration clause in the *Scott* v. *Avery* form, which makes a reference to arbitration a condition precedent, no dispute is subject to arbitration if it merely concerns a failure to pay a certified sum: *London and N. W. Railways* v. *Jones* (1915). This means that in spite of the arbitration clause the contractor can sue under the certificate and ask for summary judgment under the provisions of Order 14 of the Rules of the Supreme Court. In *Ellis Mechanical Services Ltd* v. *Wates Construction Ltd* (1976) the Court of Appeal held that retention money amounting to £52,437 out of a writ by sub-contractors claiming £187,000 against the main contractors, should be paid forthwith under Order 14, and the balance be referred to arbitration. Lord Justice Lawton said:

> 'The courts are aware of what happens in these building disputes. Cases go either to arbitration or before an Official Referee. They drag on and on and on. The cash flow is held up. In the majority of cases, because one party or other cannot wait any longer, there is some kind of compromise, very often not based on the justice of the case but on the financial situation of one of the parties. That sort of result is to be avoided, if possible. In my judgment it can be avoided if the courts make a robust approach . . . to the jurisdiction under Order 14.'

And in *Associated Bulk Carriers* v. *Koch Shipping* (1978) Lord Denning MR repeated his observations in the *Ellis* case and added: 'The court ought to give judgment for such sum as appears to the court to be indisputably due and to refer the balance to arbitration.'

He also referred to the case of *Dawnays* v. *Minter (supra)*. He said:

> 'If a set-off or counterclaim is *bona fide* and arguable up to or for a certain amount, the court may stay execution on the judgment for that amount. But in some cases, it will not even grant a stay, even when there is an arguable set-off or counterclaim, such as when the claim is on a bill of exchange . . . or for freight . . . or *for sums due on architect's certificates when they are, by the terms of the contract, expressly or impliedly payable without deduction* . . .' [author's italics].

The contractor's remedies for non-payment of an interim certificate are limited.

He can, as has been seen, sue on the certificate, in which event he will be entitled to interest from the date when payment was due: *Workman Clark & Co.* v. *Lloyd* (1908). For the reasons set out above he cannot go to arbitration over non-payment. He has no contractual right under the contract to suspend work (unlike a nominated contractor under NSC/4 or 4a clause 21.8). But he has a right under JCT 80 clause 28.1 to determine his employment under the contract in the circumstances set out in that clause [9.05].

By contrast, an architect's failure to issue interim certificates on the due date is undoubtedly a breach of contract by the employer. Article 5, which provides for arbitration, allows reference before completion in connection with 'the withholding of the architect of any certificate to which the contractor may claim to be entitled'.

But these words are not wide enough to cover what the contractor may consider under-certification, if a certificate has in fact been issued. If the certificate is on the face of it issued in accordance with the contract, the courts will not inquire whether it is in accordance with the conditions: *Kilby and Gayford* v. *Selincourt* (1973) [10.15].

6.18 Other certificates by architect

JCT 80 provides for numerous other certificates to be issued by the architect and these are dealt with specifically in the appropriate sections:

clause 17.1	on practical completion [11.01]
clause 17.4	making good defects [11.05]
clause 18.1.1 clause 18.1.3	on partial possession [11.03]
clause 22A.4	certificates for payment of insurance monies [8.07]
clause 24.1	contractor's failure to complete on due date [7.18]
clause 27.4.4	expenses incurred by employer on determination of contract by employer [9.10]
clause 30.4	retention monies [11.02]
clause 35.15.1 clause 35.16	delay by nominated sub-contractors [7.11]

6.19 The validity of the final certificate

JCT 80 clause 30.8 requires the architect to issue a final certificate. The requirements as to contents are dealt with in [11.07].

Unless specified otherwise in the Appendix, the final certificate is to be issued within three months of whichever of three specified events is the latest in time, namely:

(a) the end of the defects liability period [11.04]
(b) from 'completion of making good defects under clause 17' — curiously *not* apparently from the date of issue of the certificate of completion of making good defects referred to in JCT 80 clause 17.4 [11.04]
(c) or 'from receipt by the Architect or Quantity Surveyor of the documents referred to in clause 30.6.1.1'.

    This means 'all documents necessary for the purpose of the adjustment of the contract sum, including all documents relating to the accounts of Nominated Sub-contractors and Nominated Suppliers'.

This latter appears to be a highly nebulous and imprecise event and there appears to be no procedure for the adjudication of when event (c) takes place, apart from the opinion of the architect himself.

A final certificate issued outside the period prescribed by the contract may be void. This was so held by Judge Stabb QC in *London Borough of Merton* v. *Lowe and Pickford* (1979) and subsequently affirmed by the Court of Appeal. In this case, interestingly enough, it was the architects concerned who raised the issue that their own final certificate was invalid because issued out of time. They had issued their final certificate in May 1973, 'just over five years after they certified, in February 1968, that defects had been made good'.

Defects were known to have existed in October 1967. When these subsequently reappeared in the roof installed by the nominated subcontractors (who had gone into liquidation) the architects were sued: 'first in respect of design and/or supervision: and secondly, for issuing a final certificate and so depriving the plaintiffs of the opportunity of recovering damages from the contractor.' The contract in question was the JCT 63 Local Authorities Edition with Quantities, entered into on 8 February 1965, when the terms of JCT 63 clause 30(7) were substantially different in effect from those now in JCT 80 clause 30.9.

Judge Stabb said of that condition:

> 'The issue of a proper final certificate in accordance with the terms of clause 30 of the contract would have served as a good defence to any subsequent proceedings that the plaintiffs might have brought against the contractors for bad workmanship on the part of the sub-contractors for which the main contractors would have been liable.' [This does not apply to JCT 80 clause 30.9 nor to editions of JCT 63 from the 1976 revision: author.]
>
> 'It was contended that the final certificate was invalid and a nullity because it was given out of time . . . by reference to the Appendix, it is quite clear that the final certificate should have been given at the latest in 1969.'

His Honour referred to JCT 63 clause 30(6), which was in the same terms as JCT 80 clause 30.8 above, and held that the operative date was when the architect received the necessary documents in February 1969. 'The architect did not consider giving it until February 1970, which was clearly out of time.'

He also dealt with the contention that the final certificate was also invalid in that it did not include the sums required by JCT 63 clause 30(6)(a) — now JCT 80 clause 30.8.1 and 30.8.2.

He concluded: 'The final certificate was defective by reason of being out of time and because it did not include the sums required.' Although the final certificate was a nullity, the defendants were precluded from setting that up as a defence because that would be to rely upon their own default; in short, they were estopped by their own conduct from relying upon the nullity of the final certificate.

A final certificate may also be void if the architect who issues it is under the influence of the employer so as to cease to be impartial. In *Hickman* v. *Roberts* (1913) the architect delayed the issue of the final certificate beyond the time specified and earlier had written to the contractor: 'Had you better not call and see my clients, because on the face of their instructions to me, I cannot issue a certificate, whatever my own private opinion.'

Once an architect has issued his final certificates he can issue no others. He is, as the lawyers say, '*functus officio*' — 'out of office'.

Judge Stabb said in *Fairweather* v. *Asden* (1979):

> 'Once the architect has issued his final certificate, provided that no notice of arbitration is given by either party within the permitted time . . . the architect should be regarded as *functus officio* and is precluded from issuing thereafter any valid certificates.'

In that case, the architect had issued a final certificate on 28 July 1977. Later, on 26 March 1978, he issued a certificate under JCT 63 clause 22 that the work ought reasonably to have been completed by 10 October 1975, with the result that the employer withheld £13,300 for liquidated damages from certified sums.

In *R. M. Douglas Construction Ltd* v. *CED* (1985), Judge John Newey agreed. He was concerned with whether an architect could issue a certificate of sub-contractor's default under JCT 63 clause 27(d)(ii) (now JCT 80 clause 35.15.1) at any time. He said:

> 'Once a final certificate has been issued, or the main contract brought to an end by rescission or frustration, the architect can no longer issue a certificate under clause 27(d)(ii).'

These decisions are contrary to the view expressed in Keating: *Building Contracts*, 4th Edition, at page 345 where the learned editor wrote:

> 'It may be argued that upon the issue of the final certificate, the architect's functions under the contract are concluded but the contract does not say so and it is difficult to see the necessity for such implication.'

However, the decisions must now be accepted as binding interpretation of the JCT 63 contract and, by implication, of JCT 80.

If a certificate for payment is wrongfully withheld, the contractor can sue for the money without the certificate: *Panamona Europe* v. *Frederick Leyland & Co* (1947) (HL), and followed in *Croudace Ltd* v. *London Borough of Lambeth* (1984).

6.20 Payments for variations

The 1980 contract extends the meaning of variation by adding the following JCT 80 clause:

> 13.1.2 'the addition, alteration or omission of any obligations or restrictions imposed by the Employer in the Contract Bills in regard to:
> .2.1 access to the site or use of any specific parts of the site;
> .2.2 limitations of working space;

.2.3 limitations of working hours;
.2.4 the execution or completion of the work in any specific order; but excludes

13.1.3 nomination of a Sub-Contractor to supply and fix materials or goods or to execute work of which the measured quantities have been set out and priced by the Contractor in the Contract Bills for supply and fixing or execution by the Contractor.'

Although this appears to be no more than a minor change, it authorises the architect to make a significant difference to the manner in which the contractor carries on the work, which was formerly outside his powers.

However, this power exists only where the contract bills contain 'obligations or restrictions' on (a) access to the site, (b) limitations of working space, (c) limitation of working hours or (d) a specific order in which the work is to be carried out.

To start with, the clause excludes 'obligations or restrictions' which are in the specification or documents other than the contract bills.

Further, if there are no obligations or restrictions in the contract bills regarding, for example, hours of work, the architect cannot under this clause impose any. However, if the contract bills say there shall be no Sunday working, he can vary that by the 'addition' of an obligation that there shall be no Saturday afternoon working, or omit it to permit Sunday working.

If there is no provision as to the sequence in which work is to be executed in the bills, the architect cannot by use of this clause impose one.

This new clause has been interpreted by some writers to mean that the architect now has *carte blanche* to issue instructions to the contractor in respect of any matters specified. That is based on the argument that such instructions are an 'addition'. But one cannot 'add' to what is not there originally.

Of course, the new clause is singularly inept because JCT 80 clause 13.1.3 cannot by any stretch of the imagination come under any of the heads of clause 13.1.1 or 13.1.2.

JCT 80 clause 4.1.1 allows the contractor to make 'reasonable objection in writing' to any such instructions and a dispute regarding this is now subject to immediate arbitration under Article 5. There is no criterion available to decide what is reasonable.

Also to be treated as variations are:

clause 2.2.2.2	errors in contract bills
clause 6.1.2	architect's instructions regarding divergence between statutory requirements and the contract documents
clause 22B.2.2	restoration of damaged work
clause 22C.2.3.3	restoration of damaged work
clause 28.2.2.2	on determination by contractor 'the total value of work begun and executed but not completed at the date of determination'
clause 32.3	protective work in the event of outbreak of hostilities
clause 33.1	war damage

The implications of JCT 80 clause 13.5.5 do not seem to have been fully appreciated. It must constitute a gold mine for future contractors' claims:

> 13.5.5 'If compliance with any instruction requiring a Variation . . . substantially changes *the conditions under which any other work is executed*, then such other work shall be treated as if it had been the subject of an instruction of the Architect requiring a Variation under clause 13.2 which shall be valued in accordance with the provisions of clause 13' [author's italics].

Alter the working hours under JCT 80 clause 13.1.2.3 and all subsequent work must be substantially changed and eligible for payment not at bill rates but under the variation.

The rules for the valuation of variations include many of those used in JCT 63, but there are minor but possibly significant changes.

Additional substituted work

> 13.5.1 'To the extent that the Valuation relates to the execution of additional or substituted work which can properly be valued by measurement such work shall be measured and shall be valued in accordance with the following rules:
>
> .1.1 where the work is of similar character to, is executed under similar conditions as, and does not significantly change the quantity of, work set out in the Contract Bills the rates and prices for the work so set out shall determine the Valuation;

.1.2 where the work is of similar character to work set out in the Contract Bills but is not executed under similar conditions thereto and/or significantly changes the quantity thereof, the rates and prices for the work so set out shall be the basis for determining the valuation and the valuation shall include a fair allowance for such difference in conditions and/or quantity;

.1.3 where the work is not of similar character to work set out in the Contract Bills the work shall be valued at fair rates and prices.'

It has been suggested that the words 'similar character' should be interpreted as meaning of identical character; but this is clearly wrong. Similar means similar in the sense of 'of a like nature'. While it may be, as has been suggested by Mr Donald Keating in his *Building Contracts*, 1975 Edition, at page 316, that the words in JCT 63 clause 11(4)(b) 'similar conditions may be construed as meaning similar site and weather conditions', a change of wording in JCT 80 clause 13.5.1.1 makes it clear that 'similar conditions' applies to 'the work set out in the Contract Bills' and not to conditions in which it may have been assumed that the work would be done. Only where the bills specify the conditions under which work is to be executed can it be said that the work was executed under dissimilar conditions. This, which is the plain meaning of the words, may make nonsense of this provision since the bills rarely specify the conditions under which work is to be executed.

This part of the clause now clearly excludes work that can only be measured by time.

When the additional or substituted work is not capable of being valued by measurement, then it is provided that the variation shall be valued by:

13.5.4.1 'the prime cost of such work (calculated in accordance with the "Definition of Prime Cost of Daywork carried out under a Building Contract" issued by the Royal Institution of Chartered Surveyors and the National Federation of Building Trades Employers which was current at the Date of Tender) together with percentage additions to each section of the prime cost at the rates set out by the Contractor in the Contract Bills; or

.4.2 where the work is within the province of any specialist

> trade and the said Institution and the appropriate body representing the employers in that trade have agreed and issued a definition of prime cost of daywork, the prime cost of such work calculated in accordance with that definition which was current at the Date of Tender together with percentage additions on the prime cost at the rates set out by the Contractor in the Contract Bills.'

A footnote points out that there are three such agreements: that of the RIBA and Electrical Contractors Association; the RIBA and Electrical Contractors Association of Scotland; and the RIBA and Heating and Ventilating Contractors Association.

Omission of work set out in the contract bills is to be determined by the bill rates.

The new provision for valuation of variations therefore appears to differ significantly from JCT 63 in that:

(a) There is no allowance to be made for 'significant changes in the quantity of the work': JCT 80 clause 13.5.1.2.
(b) There is now provision for the adjustment of percentage to lump sum items in the bills: JCT 80 clause 13.5.3.2.
(c) Preliminaries in the bills of the type described in SMM6 Section B are to be adjusted: JCT 80 clause 13.5.3.3.
(d) Substantial changes as the result of a variation in the conditions under which any other work is executed are now to be treated as if they were themselves a variation: JCT clause 13.5.5.

Reference also should be made to clause 26.2.7 for possible new loss and expense claims which superficially appears to replace JCT 63 clause 11(6).

In addition, there is a new clause: JCT 80 clause 13.5.6, which is extremely mysterious in that it relates to variations which do not relate to additional work, substituted work or omitted work. Presumably it is intended to relate to the new definition in JCT 80 clause 13.1.2 [see 5.20 above] but may equally apply to errors in contract bills, statutory requirements or protective work. Anyway whatever it may mean, the contractor is to get a 'fair valuation', whatever that may mean.

6.21 The fluctuation clauses

The provisions for increases in the contract sum are now contained in three separate clauses, clauses 38, 39 and 40.

Clause 38 corresponds to the former JCT 63 (1967 printing) clause 31B and is limited to statutory and other charges and alterations.

Clause 39 is the full fluctuation clause, corresponding to JCT 63 (1967 printing) clause 31A.

Clause 40 is the Formula Method, corresponding to JCT 63 (1975 printing) clause 31F.

Unless JCT 80 clause 39 or 40 are specified in the tender documents, JCT 80 clause 38 will apply. This clause allows fluctuations relating to 'contributions, levy and tax' payable by the contractor in his capacity as an employer and, in its original form, was designed to allow the contractor relief from impositions such as Selective Employment Tax. Employers' contributions to the State pension scheme are dealt with in JCT 80 clause 38.1.8 on the basis that none is contracted out. Provision is made for 'types and rates of duty and tax . . . which at the date of tender are payable on import, purchase and sale of materials, goods, electricity and (if stated in Contract Bills) fuel'.

Clause 39 is the more extensive 'rise and fall' clause covering materials and wage rates.

Clause 40 is based on the government approved Formula Method based on monthly indices.

6.22 Fluctuations frozen on due completion date

Whichever method of calculating fluctuations is chosen, the contractor's right to receive payment for subsequent increases ceases on the date on which completion should have taken place: that is, the original contract date or such a later date as has been fixed by the architect as the result of extensions of time granted under JCT 80 clause 25. This reverses the decision in *Peak* v. *McKinney* (1970) which of course continues to apply to the fluctuation provisions in JCT 63, apart from the Formula [1.06].

JCT 80 clauses 38.4.7, 39.5.7 and 40.7.1.1 are the relevant clauses. The first two are in the same terms:

38.4.7 'no amount shall be added or deducted in the computation of the amount stated as due in an Interim Certificate or in

> the Final Certificate in respect of amounts otherwise payable to or allowable by the Contractor by virtue of clause 38.1 and .2 or 38.3 if the event (as referred to in the provisions listed in clause 38.4.1) in respect of which the payment or allowance would be made occurs after the Completion Date.'

The Formula Method one reads slightly differently, but has the same effect:

> 40.7.1.1 'If the Contractor fails to complete the Works by the Completion Date, formula adjustment of the Contract Sum under clause 40 shall be effected in all Interim Certificates issued after the aforesaid Completion Date by reference to the Index Numbers applicable to the Valuation Period in which the aforesaid Completion Date falls.
>
> .1.2 If for any reason the adjustment included in the amount certified in any Interim Certificate which is or has been issued after the aforesaid Completion Date is not in accordance with clause 40.7.1.1, such adjustment shall be corrected to comply with that clause.'

In all cases this right to freeze fluctuations is lost if JCT 80 clause 25, as printed, is amended. Therefore, if it is intended to amend that clause (e.g. by the deletion of what was formerly the optional 'starred' item, now JCT 80 clause 25.4.10.1 and .2, concerning inability to secure labour or material) it is essential to also delete clause 38.4.8.1 or clause 39.5.8.1 or 40.7.2.1 as applicable.

The right is also lost where the architect has failed to adjudicate on every written notification of an application for an extension of time under JCT 80 clause 25 within the specified time. Where this clause stands, architects should be particularly vigilant since their neglect in this respect can cost the employer dear. No longer can they afford to spike the contractor's applications for an extension of time to consider them all together at some later, much later, time.

These provisions in effect impose a monetary penalty, which may be very large, on the contractor for his breach of contract in not completing the work on the due date. It must follow that any contractor who is in this situation will want to investigate the law as to the extent to which courts of equity would hold penal clauses of this nature invalid. Support for this view will be found in *Ranger* v. *Great Western*

Railway (1854); *Public Works Commissioners* v. *Hill* (1906); *Imperial Tobacco* v. *Parslay* (1936) and in Lord Dunedin's propositions in *Dunlop Ltd* v. *New Garage Co. Ltd* (1915).

The present author confesses to sharing the view at present that those clauses which freeze fluctuations at the due date for completion may well be penalties in equity and therefore unenforceable. But an architect would be unwise to rely on that view by amending JCT 80 clause 25 or by neglecting to deal promptly with applications for extension of time.

6.23 'Work people' for the purpose of fluctuations

JCT 80 provides a definition of 'work people' in clause 38.6.3 for the purpose of that fluctuation scheme. They are 'persons whose rates of wages . . . are governed by the rules or decisions . . . of the National Joint Council for the Building Industry (NJCBI) or some other wage fixing body'. 'Wage-fixing body' is defined as a body within the meaning of the Employment Protection Act 1975, Schedule 11, paragraph 2(a). That definition is expressly said to be solely for the purposes of JCT 80 clause 38.

For the purpose of JCT 80 clause 39, there is a definition in exactly the same terms in JCT 80 clause 39.7.3 and reference is made in clause 39.1.1.1 to 'work people engaged upon and in connection with the works either on or adjacent to the site' and in clause 39.1.1.2 to 'work people directly employed by the contractor' who do not operate on the site but are engaged in the production of materials or goods intended for use on the site. In addition, for the first time, site staff who cannot be classed as 'work people' are included by JCT 80 clause 39.1.3, provided they spend not less than two whole days in any week on the site: JCT 80 clause 39.1.4.

The phrase 'employed by the contractor' is defined in the last mentioned clause as those in respect of whom PAYE is operated.

So, under the JCT 80 clause 38, there appears to be a differentiation between:

(a) 'work people' in JCT 80 clause 39.1.1.1
(b) 'work people directly employed by the contractor' in JCT 80 clause 39.1.1.2
(c) persons 'employed by the contractor' in JCT 80 clause 39.1.3.

According to the ordinary rule of interpretation a different meaning should be attached to each expression and this has led to the suggestion that JCT 80 clause 39.1.1.1 may be wide enough to include 'labour only sub-contractors' including those known popularly as 'the lump'.

This matter was considered by Mr Justice Mustill in *Murphy* v. *London Borough of Southwark* (1981) in relation to JCT 63 clause 31D(6)(c) which read:

> 'the expression "work people" means persons whose rates of wages and other emoluments . . . are governed by the rules or decisions or agreements of the NJC or some other like body for trades associated with the building industry.'

He held that 'labour only sub-contractors' were not 'work people' within that definition since their status was entirely different and their rates were not *governed* by the NJC rate. Furthermore, payments to them were not made '*in accordance with*' the NJC rules.

With respect, his decision, which was confirmed on appeal, appears to be impeccable and equally applicable to the current wording in JCT 80 clauses 38, 39 and 40. Labour only sub-contractors are not work people for the purposes of fluctuations.

6.24 Productivity and bonus payments

In *William Sindall Ltd* v. *North West Thames Regional Health Authority* (1977) increased costs by a bonus scheme, recommended by the NJCBI but not obligatory, were held by the House of Lords not to be an eligible fluctuation in wages for the purposes of JCT 63 clause 31D. JCT 80 has set out to alter that ruling by providing that the prices in the contract bills are based upon the rates of wages and other emoluments payable by the contractor in accordance with:

> 39.1.1.4 'any incentive scheme and/or productivity agreement under the provisions of Rule 1.16 or any successor to this Rule (Productivity Incentive Schemes and/or Productivity Agreements) of the Rules of the National Joint Council for the Building Industry (including the General Principles covering Incentive schemes and/or Productivity Agreements published by the aforesaid Council to which Rule 1.16 or any successor to this Rule refers) or

provisions on incentive schemes and/or productivity agreements contained in the rules or decisions of some other wage-fixing body;'

The effect of JCT 80 clause 39.1.2 appears to be that any increases in such productivity or bonus payments will henceforth be recoverable by the contractor.

The wording of JCT 80 clauses 39.1.1.4 and .5 is wide enough to cover many different incentive schemes and may therefore constitute a trap for the employer and his architect, who may not investigate carefully each separate tender in this respect.

6.25 Holiday payments

In the *London County Council* v. *Henry Boot* (1959) case, on the authority's own contract, it was held that holiday payments did not come within the term 'wages'. It is presumably for this reason that JCT 80 clause 39.1.1 uses the expression 'rates of wages and other emoluments and expenses (including holiday credits)' as did later versions of JCT 63.

6.26 Travelling costs

JCT 80 clause 39.1.5 creates new provisions regarding transport and travelling costs.

39.1.5 'The prices contained in the Contract Bills are based upon:

the transport charges referred to in a basic transport charges list submitted by the Contractor and attached to the Contract Bills and incurred by the Contractor in respect of workpeople engaged in either of the capacities referred to in clauses 39.1.1.1 and 39.1.1.2; or

upon the reimbursement of fares which will be reimbursable by the Contractor to workpeople engaged in either of the capacities referred to in clauses 39.1.1.1 and 39.1.1.2 in accordance with the rules or decisions of the National Joint Council for the Building

Industry which will be applicable to the Works and which have been promulgated at the Date of Tender or, in the case of workpeople so engaged whose rates of wages and other emoluments and expenses are governed by the rules or decisions of some wage-fixing body other than the National Joint Council for the Building Industry, in accordance with the rules or decisions of such other body which will be applicable and which have been promulgated as aforesaid.

39.1.6 If:

.6.1 the amount of transport charges referred to in the basic transport charges list is increased or decreased after the Date of Tender; or

.6.2 the reimbursement of fares is increased or decreased by reason of any alteration in the said rules or decisions promulgated after the Date of Tender or by any actual increase or decrease in fares which takes effect after the Date of Tender,

then the net amount of that increase or decrease shall, as the case may be, be paid to or allowed by the Contractor.'

In scrutinising tenders from contractors, architects should ensure that there is in fact attached a 'basic transport charges' list.

The reimbursement of fares presents much greater difficulty, since there is no provision for a contractor tendering to supply any base rates against which subsequent increases could be charted, but in the NJC Rules there are provisions regarding fares and it may well be that any charge should be taken to be based on these.

6.27 Electricity and fuels

By JCT 80 clause 39.3 the contract price is deemed to have been based on market prices of materials and goods etc. at the date of tender which is defined elsewhere as ten days before the tender is received by the employer.

These so-called 'basic prices' now include the price of electricity and where specifically so stated in the contract bills 'fuels'.

As a result, the contractor is entitled to recover increases in 'the market price of any electricity or fuels' (if in the bills) consumed on site for the execution of the works, including temporary site installations.

6.28 Notice of fluctuations

By JCT 80 clauses 38.4.1 and 39.5.1 the contractor is required to give a written notice to the architect of any increase in the matters specified in those clauses entitled him to an increase. The giving of such notice within a reasonable time is expressed to be a condition precedent to any payment.

These provisions are unlikely to be so frequently ignored as they have been in the past, as a result of the case of *John Laing Construction Ltd* v. *County and District Properties Ltd* (1982).

John Laing had entered into a JCT 63 contract which included the fluctuations clause 31D(2):

> 'any notice required to be given by the preceding sub-clause shall be given within a reasonable time after the occurrence of that to which the notice relates and the giving of a written notice shall be a condition precedent to any payment being made to the Contractor in question.'

The contractors failed to give notice within a reasonable time. Nevertheless, the quantity surveyor concerned agreed both the amount of fluctuations due and that the employer should pay them.

An arbitrator made an interim award in the form of a case stated but expressed the view that in his opinion the quantity surveyor had power under clause 31D(3) and (4) to decide what was to be paid the contractor for fluctuations.

Clause 31D(3) reads:

> 'The Quantity Surveyor and the Contractor may agree what shall be deemed for all the purposes of this contract to be the net amount payable to or allowable by the contractor in respect of the occurrence of any event . . .'

The case stated from the arbitrator raised the issue:

> 'Upon the true construction of the Contract has the Quantity Surveyor authority to make an agreement with the Claimants as to what shall be deemed to be the net amount payable to . . . in respect of the occurrence of any event such as is referred to . . . notwithstanding that the Contractor may not have given a written

notice to the Architect within a reasonable time, or at all, of the occurrence of the relevant event.'

It was held by Mr Justice Webster that the point at issue was whether the terms of this form of contract, without more, either expressly or impliedly authorised the quantity surveyor to waive non-compliance with the clause 31D(2) which required, as a condition precedent, written notice of every event.

Because of the use of the word 'deemed' the words of clause 31D(3) were apt to describe *quantum* only but not the employer's liability to pay. The emphasis was on the 'net amount' rather than on the words 'payable or allowable'.

The quantity surveyor's 'functions and authority under the contract are confined to measuring and quantifying'. Nowhere in the contract was he given authority to determine any liability, or liability to make any payment or allowance.

The words in clause 11(4) 'unless otherwise agreed' do not mean by or with the quantity surveyor.

It is probable that JCT 80 clauses 38.4.3, 39.5.3 and 40.5.3 confer greater powers on the quantity surveyor than do JCT 63 clauses 31F and 31D(3); but it is debatable.

6.29 Sub-contractors and fluctuations

The provisions of the main contract regarding fluctuations are to be extended to all work sub-let by the contractor to a domestic sub-contractor: JCT 80 clauses 38.3.1 and 39.4.1.

In the case of nominated sub-contractors, NSC/4 and 4a makes provision similar to the main contract in NSC/4 or 4a clause 35 (equals JCT 80 clause 37), clause 36 (equals JCT 80 clause 28) and clause 37 (equals JCT 80 clause 40).

The once vexed question of contractors' cash discount on fluctuations has been resolved, it is claimed, by NSC/4 or 4a clause 21.10.2.11 which includes an item to be included in the adjustment of the sub-contract sum by the addition of 'an amount equal to one thirty-ninth of the amounts' specified in clauses 21.10.2.5, 21.10.2.8 and 21.10.29 and amongst those so specified are the payments under an appropriate fluctuations clause: NSC/4 or 4a clause 21.10.29.

The freezing of fluctuations on the part of a nominated sub-contractor is not governed by whether or not the main contractor has passed his completion date. They are governed by whether or not the sub-contractor has passed his: NSC/4 or 4a clauses 35.4.7, 36.5.7 and 37.7.1.

A nominated sub-contractor who sub-sub-contracts work is under a contractual obligation to include similar terms in all such sub-sub-contracts: NSC/4 or 4a clauses 35.3 and 36.4.

6.30 Formula Rules: JCT 80 clause 40

The Formula Rules originated with a report of the National Economic Development Council for the Building Industry in 1969. The principle was applied to the ICE contracts in 1973, GC/Works/1 in 1974 and the JCT contracts in 1975 with JCT 63 clause 31F. The current edition is the 'Formula Rules for use with the Standard Form of Building Contract 1977 as revised in 1980'.

The formula system requires that the bills of quantities should allocate work to the appropriate work category corresponding to the indices. A base rate is thus established for the tender date.

As has been seen a valuation is required for each monthly interim payment as the work progresses and these are then adjusted in accordance with the published indices and the terms of the appropriate contract provision. For details, reference should be made to the NEDC documents *Guide to Practical Application of the Formula* and *Description of the Indices* and to the PSA's monthly *Bulletin of Construction Indices*.

Chapter 7

Liquidated damages and extensions of time

7.01 Provisions regarding liquidated damages

The RIBA and JCT contracts for many years have provided for a sum to be named as liquidated damages for the failure of the contractor to complete the works on time. Fixed sums of this nature, payable on a breach of contract, are usually referred to in the construction industry as 'the penalty clause'. They are not, however, normally what the courts of equity would term a penalty, which is a sum so extravagant as to bear no relationship to any loss the injured party might suffer from such a breach.

The distinction between liquidated damages, penalties and limited damages clauses is adequately dealt with in most books on contract, including the present author's book *Commercial Law* and it is not proposed to discuss them here.

Some misconceptions prevalent in the construction industry should however be dealt with. The sum named as liquidated damages in the contract is recoverable whether or not the employer can prove that he has in fact suffered any loss or damage as the result of the breach: *Crux* v. *Aldred* (1866). 'The courts of equity never undertook to serve as a general adjuster of men's bargains,' said Lord Radcliffe in *Bridge* v. *Campbell Discount Co. Ltd* (1962) and if the contractor has agreed to pay this sum, if he breaks the contract by not completing on time, the courts will hold him to his agreement. The fact that damages when the contract was made are difficult to quantify or estimate, as in the case of a church, does not prevent a sum named in the contract being liquidated damages: *Fletcher* v. *Dycke* (1787).

Secondly, some have been misled by the generalised observations of Lord Justice Salmon as he then was, in *Peak* v. *McKinney* (1970) that 'liquidated damages and extension of time clauses in printed forms of contracts must be construed *contra proferentem*' the owner, for whose exclusive benefit, his lordship held, they exist. But he was dealing with a contract which was not a JCT one but was devised by Liverpool Corporation and, according to his lordship, 'if a prize were to be offered for the form of a building contract which contained the most one-sided, obscurely and ineptly drafted clauses in the United Kingdom, the claim of this contract could hardly be ignored, even if the RIBA form of contract was among the competitors'. Clearly the whole contract was the employer's one and was to be construed *contra proferentem* him.

It is submitted that the JCT liquidated damages clause and extensions of time provisions are not to be construed *contra proferentem* either party [1.07]. They are there for the benefit of both. It is a wholly incorrect approach to consider each clause and attribute each to one or other party. The contract must be taken as a whole: *Terson Ltd* v. *Stevenage Development Corporation* (1963).

However, in *Bramall and Ogden Ltd* v. *Sheffield City Council* (1983), Judge Hawser appears to have been misled into following the *dicta* of Lord Justice Salmon in *Peak* v. *McKinney* (*supra*). It would appear from his judgment that none of the relevant authorities (there are many) was quoted to him.

In the nineteenth century all liquidated damages clauses were considered as being exclusively in the employer's interests. More recently it has been recognised by the courts that they can be in the interests of both parties especially where they are contracts agreed by representatives of both employers and contractors.

The case appears to be manifestly wrong in law on another point as well. The JCT 63 contract provided that liquidated damages could be recovered 'at the rate of £20 per week for each uncompleted dwelling', a well-established method of calculating the amount of liquidated damages in local authority housing contracts. The judge mistakenly took this to mean it was provision for sectional completion. Possibly it was not appealed because the local authority concerned could recover more by way of general damages than those provided for by the contract; possibly because they had forfeited their right to liquidated damages because variations had been ordered after the due date for completion [7.21].

7.02 Time at large and loss of right to liquidated damages

However, two propositions can be advanced with confidence. The employer cannot rely upon a liquidated damages clause when he himself has been responsible in part or wholly for the failure of the contractor to complete on time. The purpose of JCT 80 clause 25 is to enable the employer to recover liquidated damages. If not there, he would forfeit right to liquidated damages if any part of the delay was his fault: *Holme* v. *Guppy* (1838):

> 'Later cases show that if completion by the specified date was prevented by the fault of the employer, he can recover no liquidated damages unless there is a cause providing for extension of time in the event of delay caused by him': Staughton J. in *Astilleros Canarios SA* v. *Cape Hatteras Shipping* (1981).

If part of the reason why the contractor cannot complete on time is the employer's fault, time may be 'at large', i.e. there will be no fixed date for completion, only for completion 'in a reasonable time'. And the employer will be put to proof of what damage he actually suffered if it is not completed in a reasonable time.

Referring to JCT 63 in the Court of Appeal in *Percy Bilton* v. *Greater London Council* (1981) Sir David Cairns said:

> 'There are several paragraphs in clause 23 where the right to apply for extension of time is based on the fault of the employer. 23(f) is the clearest of such cases and it would be inconsistent with its terms to hold that a delay by the employer in giving any necessary instructions could cause time to be at large.'

But that may be lost if the architect, with such power to extend, wrongfully fails or refuses to exercise his powers: *Peak* v. *McKinney* (1970).

7.03 Employer's failure to renominate

In *Percy Bilton* v. *Greater London Council* (1980) the senior Official Referee, Judge Stabb, held that the liquidated damages clause becomes entirely inoperative if a nominated sub-contractor becomes insolvent or withdraws under any other situation from the site and the contractor is thereby delayed.

The plaintiffs were many weeks behind with their own work and the nominated sub-contractor for mechanical services was not able, for this reason, to start work for 22 weeks. Thirty-six weeks later the nominated sub-contractors withdrew from the site and went into liquidation on 31 July 1978. The same day another contractor for mechanical services was engaged on a daywork basis. By 14 September 1978 the mechanical services contractor employed on a daywork basis became the nominated sub-contractor. The architect extended time for completion under JCT 63 clause 23(f) for this delay in renomination.

On 4 February 1980 the architect issued a certificate under JCT 63 clause 22 that the contract ought reasonably to have been completed by 1 February 1980. Thereafter the GLC began to deduct liquidated damages amounting to £97,543 in all, from the sums certified in interim certificates.

Judge Stabb held:

(1) The certificate under JCT 63 clause 22 was invalid since the architect has no power under JCT 63 clause 23(f) to extend time for delay caused to the contractor by the employer's failure to renominate.
(2) As the result of delay caused to the contractor by the renomination of a sub-contractor, the liquidated damages clause was ineffective, 'time was at large', and the employer could only recover such general damages as he could prove he had suffered for failure to complete 'in a reasonable time'. He said:

'Only an employer, through his architect, is entitled to nominate a sub-contractor and if that nominated sub-contractor repudiates or ceases work before the sub-contract is completed, it is the duty of the employer to renominate a second sub-contractor to complete the work if the contract works are indeed to be completed: *Bickerton* v. *North West Metropolitan Hospital Board* (1970).

If the employer fails to renominate a sub-contractor, he is in breach of contract . . .

Nomination and renomination of sub-contractors are the responsibility of the employer and . . . it is the responsibility of the employer to nominate or renominate in a manner which does not impede or delay the main contractor in executing or completing the works.

[As to JCT 63 clause 22, he said:] The date by which the architect can certify that the works ought reasonably to have been completed must be the date for completion stated in the contract

> or within any extended time granted by virtue of any of the provisions of clause 23.
>
> Any certificate issued under clause 22 purporting to extend the time for completion for a reason not covered by clause 23 must be invalid and, accordingly, there remains no basis for a claim for liquidated damages.'

He quoted Lord Justice Salmon, as he then was, in the *Peak* case:

> 'If the failure to complete on time is due to the fault of both the employer and the contractor, in my view, the clause does not bite. I cannot see how, in the ordinary course, the employer can insist on compliance with a condition if it is partly his own fault that it cannot be fulfilled: *Wells* v. *Army & Navy Co-operative Society Ltd* (1902); *Amalgamated Building Contractors* v. *Waltham Urban District Council* (1952); *Holme* v. *Guppy* (1838).
>
> Unless the contract expresses a contrary intention, the employer, in the circumstances postulated, is left to his ordinary remedy; that is to say, to recover such damages as he can prove flow from the contractor's breach.'

The judgment in the *Percy Bilton* case was set aside by the Court of Appeal in 1981.

Sir David Cairns, in the Court of Appeal, said:

> 'None of the cases support the proposition that the mere repudiation of the sub-contract by a nominated sub-contractor is to be regarded as a fault or breach of contract on the part of the employer and it appears to me quite unreasonable to so describe it.
>
> The duty to renominate . . . must . . . be a duty to renominate within a reasonable time. The delay caused by the departure of [the sub-contractors] . . . was a delay not within any of the provisions of clause 23. The date for completion remained unaffected by that delay.'

He then turned to the delay caused by failure to nominate for several months. That delay, he said, 'seems to be quite separate and clearly within the provisions of clause 23(f)'. And later: 'If the architect failed to give adequate extension, the matter could be taken to arbitration under clause 35.'

He therefore concluded that the GLC was entitled to liquidated damages from the extended time that the architect had allowed. Lord Justice Stephenson and Lord Justice Dunn agreed with him.

The case went to the House of Lords.

The House of Lords noted that the sub-contractor's employment had come to an end by notice given by a receiver that labour would be withdrawn and this, it was held, constituted a repudiation of the sub-contract which Percy Bilton had accepted.

It was common ground that the delay caused to the contractor fell into two parts:

(a) part arising from the withdrawal of the sub-contractor
(b) part from the failure of the GLC to nominate a replacement sub-contractor with reasonable promptness.

For the latter period, for which the GLC accepted responsibility, Percy Bilton had been given an extension of 14 weeks.

A passage in the judgment of Lord Justice Stephenson in the Court of Appeal was expressly approved:

> 'Insofar as delay was caused by the departure of [the sub-contrac tor] . . . it was a delay which was not within the provision of clause 23.
>
> Therefore the plaintiff was not entitled to any extension of time in respect of it, with the result not that time became at large but that . . . the date for completion remained unaffected.'

In the House of Lords, Lord Fraser said:

> 'The true position is . . . correctly stated in the following propositions:
>
> (1) The general rule is that the main contractor is bound to complete the work by the date for completion stated in the contract. If he fails to do so, he will be liable for liquidated damages to the employer.
>
> (2) That is subject to the exception that the employer is not entitled to liquidated damages if by his acts or omissions he has prevented the main contractor from completing his work by completion date — see for example, *Holme* v. *Guppy* (1838), and *Wells* v. *Army and Navy Co-operative Society* (1902).
>
> (3) These general rules may be amended by the express terms of the contract.
>
> (4) In this case, the express terms of clause 23 of the contract do affect the general rule. For example, where completion is delayed

'(a) by *force majeure*, or (b) by reason of any exceptionally inclement weather', the architect is bound to make a fair and reasonable extension of time for completion of the work. Without that express provision, the main contractor would be left to take the risk of delay caused by *force majeure* or exceptionally inclement weather under the general rule.
(5) Withdrawal of a nominated sub-contractor is not caused by the fault of the employer, nor is it covered by any of the express provisions of clause 23. Paragraph (g) of clause 23 expressly applies to "delay" on the part of a nominated sub-contractor but such "delay" does not include complete withdrawal . . .
(6) Accordingly, withdrawal falls under the general rule, and the main contractor takes the risk of any delay directly caused thereby.
(7) Delay by the employer in making the timeous nomination of a new sub-contractor is within the express terms of clause 23(f) and the main contractor, the appellant, was entitled to an extension of time to cover that delay. Such an extension has been given.'

The exact terms of JCT 63 clause 23(f) were:

'by reason of the Contractor not having received in due time necessary instructions, drawings, details or levels from the Architect for which he specifically applied in writing on a date which having regard to the Date of Completion stated in the appendix to these Conditions or to any extension of time then fixed under this clause or clause 33(1)(c) of these Conditions was neither unreasonably distant from nor unreasonably close to the date on which it was necessary for him to receive the same . . .'

Nowhere in the JCT 63 contract does it suggest that the main contractor is under an obligation to apply to the architect *in writing* for a re-nomination. In fact, it was clearly held by the House of Lords in *T. A. Bickerton* v. *North West Metropolitan Regional Hospital Board* (1970) that the obligation was on the employer to provide a replacement sub-contractor. But in this case Percy Bilton did apply *in writing* for the architect's instructions. What if they had not done so? Had the architect then any power under clause 23(g) to extend time for the employer's unreasonable delay in re-nomination? If he had no such power under the contract, time must have been at large, since it was common ground that in regard to the second period of delay, the employer was at fault. Did that one letter cost the contractor what is believed to be £2 million in liquidated damages?

7.04 Employer's failure to give possession of site

A term is to be implied in every building contract that the employer will give possession of the site to the contractor in sufficient time for him to complete his operations by the contractual date: *Freeman* v. *Hensler* (1900). 'There is an implied undertaking on the part of the building owner who has contracted for buildings to be placed . . . on his land, that he will hand over the land for the purpose of allowing [the contractor] to do that which he has bound himself to do,' said Lord Collier CJ. This is a term which will be implied as the result of *The Moorcock* doctrine [2.06].

JCT 80 clause 23.1 makes a specific provision for the contractor to be given possession on the date specified in the Appendix.

It follows that failure to do so on that date is a breach of the express and implied terms of the contract by the employer. At the very least the contractor will be entitled to damages for any loss suffered by this breach: *London Borough of Hounslow* v. *Twickenham Gardens* (1971).

If there is no provision in the contract for the architect to grant an extension of time for completion, it will mean that the date for completion will be at large and the employer will forfeit his right to liquidated damages.

In his book *Building Contracts* (2nd Edition, page 343), Donald Keating took the view that under JCT 63 failure to give possession was covered by the clause (now JCT 80 clause 23.2) entitling the architect to postpone 'any work to be executed under the provisions of this contract' and that the architect could then extend time for completion under what is now JCT 80 clause 25.4.5. But in fact a great deal more is involved in giving the contractor possession of the site than just allowing him to execute work and JCT 80 clause 23.2 is not wide enough to cover failure to do so: *Roberts* v. *Bury Commissioners* (1880); *London Borough of Hounslow* v. *Twickenham Gardens* (1971).

It follows that there is now no 'relevant event' [see 6.06] which entitles the architect to extend completion time for failure to give possession and unless he has other power to do so, the liquidated damages clause may be worthless for reasons set out in [7.02] and time may be at large.

This situation appears now to have been accepted by the Joint Contracts Tribunal, for the Intermediate Form IFC 84 contains a provision entitling the architect to postpone giving possession on the due date, subject to conditions.

It is understood that a similar clause is in preparation as an amendment to JCT 80.

It has been suggested that the situation may possibly be saved by JCT 80 clause 25.3 where the architect is under an obligation to 'fix a Completion Date later than that previously fixed if such later Completion Date is fair and reasonable': see [7.01].

This is a re-run of the argument that the words in JCT 63 clause 22, which required an architect to certify in writing that 'in his opinion the same ought reasonably so to have been completed', gave a kind of equitable jurisdiction to the architect whereby he could extend time for reasons other than those set out in JCT 63 clause 23 and for periods other than those he had previously allowed.

That controversy has been settled by the decision in the *Bilton* v. *GLC* case *supra* and there is no reason why the same argument should not apply to the new wording in JCT 80 clause 25.3.

Unhappily, failure to give possession on the due date is not uncommon, often for good reason, e.g. that a demolition contractor has not finished in time as happened in *Neville (Sunblest) Ltd* v. *Press* (1981). It is desirable that the contract should be amended to cover this possibility, if it is likely. The JCT contract is not as immutable as the law of the Medes and Persians and the terms can always be varied by agreement between the parties even after execution. But the architect has no express, implied or ostensible power to do so without the authority of his principal.

7.05 Time for completion

The date for completion is defined in JCT 80 clause 1.3 as 'the date fixed and stated in the Appendix'. The date for possession is defined as 'the date stated in the Appendix under the reference to clause 23.1'.

The provisions of the contract regarding extensions of time have been extended and are now in JCT 80 clause 25. This clause is solely for the purpose of granting the contractor a deferment of the completion date. It has no bearing whatsoever on any claim for increased expenses or losses. The provisions of JCT 80 clause 26 are discussed subsequently [10.03] to emphasise that they are totally unrelated to clause 25.

The *JCT Guide*, issued in 1980 with the new contract, attempts to describe the effect of changes. On page 11 in paragraph 14.1 it reads:

> 'The Contractor must now give notice, particulars and estimates of likely delay to completion in respect of any matter which the Contractor has identified as a Relevant Event . . . and he must ensure that information is kept reasonably up-to-date.'

This suggests that the architect has no power to grant an extension in the absence of such a notice. That is not what the contract says in JCT 80 clause 25.3.3, which provides for a review of the completion date not later than the expiry of 12 weeks from the date of practical completion 'whether the Relevant Event has been specifically notified by the Contractor or not'. The only consequence of the contractor failing to give notice of a relevant event is to postpone the time when the architect has to decide. In any event, as a matter of law, the employer could not deprive the contractor of his right to an extension of time for failure to give notice without forfeiting his right to liquidated damages.

The *JCT Guide* at page 11, paragraph 14.4 also states:

> 'The Architect is permitted after his first decision on extending time to take account of any Variations requiring an omission which have been issued since the date of his last decision on extending the Completion Date.'

In fact, the architect can take account of omissions when adjudicating upon the very first extension application: JCT 80 clause 25.3.1 which reads in part:

> 'If, in the opinion of the Architect, *upon receipt of any notice* . . . under clause 25.2.1.1 . . . the architect shall . . . state [25.3.1.4] the extent, if any to which he has had regard to any instruction . . . requiring . . . the omission of any work issued since the fixing of the previous Completion Date' [author's italics].

The completion date is defined in JCT 80 clause 1.3 as: 'the Date for Completion as fixed and stated in the Appendix or any later date fixed under either clause 25 or 33.1.3.'

It follows that there is always a 'previous Completion Date' from the moment the contract is signed.

JCT 80 clause 25.3.6 prohibits an architect from fixing an earlier completion date than that named in the Appendix, *whatever* omissions are made.

Subject to that overriding provision, he is authorised, *on his own initiative*, whenever he issues a variation order for an omission to reduce a previously granted extension: JCT 80 clause 25.3.2.

> 'After the first exercise by the Architect of his duty under clause 25.3.1 the Architect may in writing fix a Completion Date earlier than that previously fixed under clause 25 if in his opinion the fixing of such earlier Completion Date is fair and reasonable having regard to any instructions of the Architect requiring as a Variation the omission of any work issued under clause 13.2 where such issue is after the last occasion on which the Architect fixed a new Completion Date.'

The authors of the *JCT Guide* appear to have overlooked the existence of JCT 80 clause 25.3.1.4.

7.06 Relevant events: JCT 80 clause 25.4

This is the term now used for matters for which the architect is entitled to retard the date for completion by the contractor.

These can conveniently be categorised as:

- Those which result from acts or omissions of the employer or his architect or others acting on his behalf. This is perhaps the most important category since without such clauses the employer would be liable to forfeit his right to liquidated damages, e.g. by ordering variations as in *Dodd* v. *Churton* (*supra*). It will not escape notice that under this clause the architect is required to adjudicate upon matters which may be his own default: JCT 80 clauses 25.4.5, 25.4.6, 25.4.8 and 25.4.12.
- Those where the contractor is delayed in completion as the result of the acts or omissions of persons other than the contractor or the employer or those for whom he is responsible: JCT 80 clauses 25.4.7, 25.4.9, 25.4.11 and 25.4.4.
- Events which are outside the control of either of the contracting parties: JCT 80 clauses 25.4.1, 25.4.2, 25.4.3, 25.4.4, 25.4.9, 25.4.10 (and possibly 25.4.11).

It will be observed that the policy of this section is not to grant extensions of time (apart possibly for some events in the clause 22 perils) for which the contractor himself is wholly or partly responsible, and there is express and overriding provision qualifying all applications for extensions of time:

> 25.3.4 'Provided always
> .4.1 the Contractor shall use constantly his best endeavours to prevent delay in the progress of the work, *howsoever caused* and to prevent completion of the Works being delayed or further delayed beyond Completion Date;
> .4.2 the Contractor shall do all that may reasonably be required *to the satisfaction of the Architect to proceed* with the Works' [author's italics].

Other clauses qualify the contractor's right to receive an extension of time:

> 25.4.7 '. . . which the Contractor has taken all *practicable* steps to avoid or reduce;'
> 25.4.10.1 'for reasons beyond his control and which he could not reasonably have foreseen . . .' [author's italics].

Architects are, however, strangely reluctant to refuse extensions on any of these grounds.

What are 'best endeavours'?

It is a widespread opinion in the construction industry that clause 25.3.4.1 requires the contractor only to use reasonable efforts to prevent delay in the progress of the work. For example in *Contract Documentation for Contractors* (1985) Vincent Powell-Smith and John Sims state:

> 'Some architects assume that this amounts to an obligation to mitigate a delay which has already occurred or which is unavoidable. When the delay is caused by a "relevant event" this is not so, nor can the architect require you to spend additional money, for instance by accelerating to meet the obligation.'

They provide a specimen letter dismissing the architect's requirement for them to accelerate the work to make up for time lost by exceptionally inclement weather.

With respect to the experienced authors, that is not what the contract says. Clause 25.3.4.2 is the only sub-clause which uses the words 'reasonably . . . required'. It has nothing whatsoever to do with clause 25.3.4.1, but because it exists, the word 'reasonable' cannot be written into 25.3.4.1.

The clause 25.3.4.1 demands of the contractor not *his* reasonable or reasonably economic efforts but his best efforts. If that consists of increasing the labour force, working overtime or bringing in more plant, the contract requires him to do it, even if it costs him extra expense for which he will not be reimbursed.

The present author quite agrees that the architect has no power under the contract to demand that the contractor accelerate his work to 'prevent completion of the works being delayed . . .' beyond completion date.

But what the architect *is* entitled to say is that the contract entitles him to refuse to extend time for a relevant event because the contractor has not used his *best* endeavours to prevent delay.

This is a standard form contract in the formulation of which the contractors' representatives have a substantial, if not paramount, influence. Had they insisted, no doubt the JCT would have inserted the words 'the contractor shall use reasonable endeavours to prevent delay in the progress of the works'.

That is not what it says in clause 25.3.4.1. He undertakes to use his *best endeavours*. In short, if it is within his power, *by any means whatsoever*, to prevent delay to the progress of the works, he has undertaken to exercise that power. And, if he has not done so, the architect is fully entitled to deny him an extension of time.

This is what the contract plainly says.

Any architect who allows a contractor, who has not exercised his 'best endeavours', an extension of time is plainly liable to his principal. The provision may be onerous, it may even be unreasonable; but this is what the contractor has specifically agreed. 'Best' means 'best'; it does not mean any steps which will not cause the contractor additional expense.

7.07 Compliance with architect's instructions: JCT 80 clause 25.4.5

Only certain architect's instructions are specified as coming within this category, namely:

clause 2.3: Discrepancies in contract bills etc.

clause 13.2: Variations
clause 13.3: Expenditure of provisional sums in contract bills and sub-contracts
clause 23.2: Postponement of contract works
clause 34: Antiquities
clause 35: Nominated sub-contractors
clause 36: Nominated suppliers

This clause is more extensive than JCT 63 clause 23(e) and it would appear that the question which arose in *Percy Bilton Ltd* v. *Greater London Council* (1981) as to whether there was power under JCT 63 to extend time for delay in renomination of a sub-contractor would now come under this clause and not under JCT 80 clause 25.4.6 which is the replication of JCT 63 clause 23(f). JCT 80 clause 35.24 deals specifically with circumstances where renomination is necessary; therefore the architect is entitled, and indeed obliged, to issue instructions for renomination. There can be no doubt therefore that under JCT 80 an architect has power to extend the contractor's time if a nominated sub-contractor goes into liquidation and delay in renomination results.

Delay caused by architect's instructions to open up for inspection is included in JCT 80 clause 25.4.5.2:

> 'in regard to the opening up for inspection of any work covered up or the testing of any of the work, materials or goods in accordance with clause 8.3 (including making good in consequence of such opening up or testing) unless the inspection or test showed that the work, materials or goods were not in accordance with this contract.'

7.08 Instructions, drawings not in time: JCT 80 clause 25.4.6

> 25.4.6 'the Contractor not having received in due time necessary instructions, drawings, details or levels from the Architect for which he specifically applied in writing provided that such application was made on a date which having regard to the Completion Date was neither unreasonably distant from nor unreasonably close to the date on which it was necessary for him to receive the same;'

On the face of it the architect is only empowered to extend time if the contractor has 'specifically applied in writing' within the period specified.

This has always raised very difficult questions, which the JCT appear to have ignored. Failure by the architect to produce these documents etc. in time for the contractor to do the work is undoubtedly a breach of the express and implied terms of the contract. If the architect has no power, without application in writing in due time by the contractor, to grant an extension of time, such breaches may result in the employer losing his right to liquidated damages [6.02]. So what is an architect to do where the contractor has either (a) made no written application at all for the necessary instructions; or (b) has made a belated one?

It would appear that it is not a relevant event and he has no power to extend time. However, it is perhaps possible that he may subsequently have power to entertain it on his review [6.20] but this will turn on the interpretation to be placed on words in JCT 80 clause 25.3.3 [7.04].

It can also be argued that a contractor who does not give notice in writing is the author of his own wrong and not entitled to any relief.

7.09 Work not forming part of the contract: JCT 80 clause 25.4.8

Clause 25.4.8 has been rewritten since it appeared as JCT 63 clause 23(h): 'delay on the part of artists, tradesmen or others engaged by the Employer in executing work not forming part of this contract.'

This was known as the 'Epstein clause' because it appeared in the JCT contract after that sculptor had delayed a contractor by failing to produce his work on time and it was found there was then no provision to extend time for such an event. In *Henry Boot Construction Ltd* v. *Central Lancashire New Town Development Corporation* (1980) Judge Fay QC said of this clause that the words 'or others' were not to be construed *ejusdem generis* and that delay caused by statutory undertakers engaged by an employer under contract (as an arbitrator had found as a fact) came under this clause and the similarly worded JCT 63 clause 24(d).

It will be seen that, apparently by accident, the words 'delay on the part of' have been lost in the revision. Clause 26.2.4.1 now reads:

> 'the execution of work not forming part of this Contract by the Employer himself or by persons employed or otherwise engaged

> by the Employer as referred to in clause 29 or the failure to execute such work.'

JCT 80 clause 29 obliges the contractor to permit the execution of such work provided that reference is made to it in the contract bills; and that if no such reference is made 'the Employer may with consent of the Contractor (which consent shall not be unreasonably withheld) arrange for the execution of such work'.

The decision in *Henry Boot* v. *Central Lancashire New Town Development Corporation* is irrelevant to the new clauses save for the explanation given by Judge Fay as to the meaning of the words 'not forming part of this contract' which appeared in JCT 63 clause 23(h) and now appears in JCT 80 clause 25.4.8, 26.2.4 and clause 29.

In that case, JCT 63 (July 1975 revision) provided by Article 1 that the contractors should execute the work shown on the contract drawings and described or referred to in the contract bills etc. and by Article 2 that the employer would pay them the sum of £2,765,716 'hereinafter referred to as the "Contract Sum"'. The work by three statutory undertakers was contained in the contract bills under 'Direct Payments: Local Authorities and Public Undertakings — Provide the following sums for work to be executed . . . Electric Main and Sub-station £28,500.'

There were similar provisions regarding water mains, gas mains and electrical connection to the street lighting system. There was also the further provision:

> 'The amounts included for Works to be executed by Local Authorities or public undertaking are to be expended under direct order of the Central Lancashire Development Corporation'

and

> 'Liaison with public bodies — The Employer intends to give permission for or instruct the following public bodies to carry out works during the progress of the Works: Local Authority; Highway Authority; Water Board; Gas Board; Electricity Board; Post Office. Afford all reasonable facilities to these bodies, and give ample notice when their work may proceed without interruption and in accordance with the programme.'

The sums relating to work by these bodies were included in 'the Contract Sum of £2,765,716'. Provision was made for the deduction of the specified sums in the final account.

His Honour said:

> 'Just why this remarkable device has been adopted of putting this work in with one hand and taking it out with the other neither side in their hearing has satisfactorily explained . . .
>
> So I now reach the position that by Article 1 of the contract the contractor binds himself to carry out works which, under the bills of quantities, incorporated into the contract, he is told not to do, and he is told moreover that others will do.
>
> And under Article 2 he is to be paid a total sum, including part for work which he is not to do and which the bills of quantities provide that he shall not receive.
>
> Does or does not this work form part of the contract?'

His Honour considered the implications of what is now JCT 80 clause 2.2.1, to the effect that nothing in the bills should override or modify the printed Articles or Conditions or Appendix and decided, following *English Industrial Estates* v. *Wimpey* (1973), that he was entitled to look at the bills to 'follow what was going on'. He concluded:

> 'For some purposes the work does form a part, literally, of the contract; but for other purposes it does not.
>
> It is not work which the employer can require the contractor to do. All that he can require is that the contractor affords attendance etc. on those who do the work . . . [and that] I take the pragmatic view that the relevant work is work not forming part of the contract.'

If that be right, delay that is caused by a statutory undertaker, provided it is not that caused in the carrying out of work 'in pursuance of its statutory obligation' will come within JCT 80 clause 25.4.8.1 and clause 26.2.4 and therefore the employer will be liable to pay damages to the contractor without hope of redress from the statutory undertakers.

All this follows from the finding of fact by the arbitrator, by which the judge was bound, since it came before him only on an award by way of case stated, that 'the statutory undertakers carried out their work in pursuance of a contract with the employers'. It does not appear from the award whether the legal position of statutory undertakers [5.16] was properly explored before the arbitrator.

It would appear from evidence that was not before the judge that the statutory undertakers concerned were in fact doing no more than carrying out their statutory obligations and were not in any contractual relationship with the employers, so that the arbitrator's finding of fact was wrong.

The ingenious argument that the work was not 'forming part of this contract' when it was priced in the contract sum and included in the contract bills is unconvincing. The expression 'artists, tradesmen and others' also seems an inappropriate expression to apply to the local electricity board, water board and gas board. No doubt from time to time they are termed 'artists', but always with an adjectival qualification.

Architects should therefore be very reluctant to extend time under JCT 80 clause 25.4.8 without satisfactory proof from the contractor (a) that the work was not being carried out by the undertaker pursuant to his statutory obligations, and (b) that in fact the undertaker was a person 'employed or otherwise engaged by' the employer within the meaning of JCT 80 clause 29.

Failure of the employer to supply materials and goods which he has agreed to provide is a new ground for extension of time: JCT 80 clause 25.4.8.2, but does not entitle the contractor to any loss or expense claim under JCT 80 clause 26.

7.10 Failure to give ingress or egress

A new clause 26.2.6 reads:

> 'failure of the Employer to give in due time ingress to or egress from the site of the Works, or any part thereof through or over any land, buildings, way or passage adjoining or connected with the site and in the possession and control of the Employer, in accordance with the Contract Bills and/or the Contract Drawings, after receipt by the Architect of such notice, if any, as the Contractor is required to give, or failure of the Employer to give such ingress or egress as otherwise agreed between the Architect and the Contractor.'

The same is 'a matter' entitling the contractor to direct loss and/or expense under JCT 80 clause 26 [10.03].

As has been pointed out earlier [7.04] this clause in no way entitles the contractor to an extension of time for the employer's failure to give him possession of site.

It is clearly an implied term of all construction contracts that the employer will do nothing to impede the contractor's free access to the site for the purpose of executing the work he has contracted to do. Accordingly, for the employer to impede access will be a breach of contract which will entitle the contractor to damages and in the absence of this provision for extension might also disentitle the employer to recover liquidated damages.

It is not, however, a breach of contract where access is impeded by other parties. In *LRE Engineering Services Ltd* v. *Otto Simon Cares Ltd* (1981), the plaintiffs were sub-contractors to the defendant for work at steelworks at Port Talbot. Clause 2a of the sub-contract terms contained an express provision for access and in particular clause 24(1) provided that 'access to and the possession of the site shall be afforded to LRE by Otto Simon in proper time for the completion of the work'.

The sub-contractor had carried out a substantial proportion of the work when a steel strike broke out on 3 January 1980 which continued until 8 April. Picketing at the works prevented LRE completing the work and when the strike was over they incurred expenses of £370,000 in putting the work back into the condition in which they had left it. LRE sued for damages for breach of clause 24(1). Mr Justice Goff held that there had been no breach of contract on the part of Otto Simon. It was the activities of third parties which prevented LRE gaining access and that did not amount to a breach of the terms of the contract.

Those principles apply with full force to applications for extensions of time under JCT 80 25.4.12 and loss and expense under 26.2.6.

A contractor is not entitled to an extension of time unless 'ingress or egress from the site' has been prevented by the Employer himself or somebody for whom he is vicariously liable.

Moreover, the right to an extension is also limited to lands which are 'in the possession and control of the employer' and these alone. If, for example, the obstruction takes place on the highway (including any verges) outside the site, the contractor is not entitled to an extension of time, let alone damages, either at common law or under clause 26. Therefore, if the local highway authority were to close up a road, which was the only access to the site, as they did in *National Carriers Ltd* v. *Panalpina (Northern) Ltd* (1980) because of the unsafe condition of a warehouse on it, and keep it closed for say 20 months, the contractor would be entitled to no relief. But, as was said in that case, if the suspension is for a sufficiently long period, it might amount to a frustration of the contract. And it might possibly be *force majeure* under clause 28.1.3.1, though probably not under clause 25.4.1.

It is clear therefore that neither JCT 80 25.4.12 nor 26.2.6 imposes a strict liability on the employer to ensure access to the contractor. The contractor will not be entitled to an extension of time if, as happened in Oxfordshire, gypsies parked their caravans so as to block the only access to the site. That was not a 'failure of the Employer' and the land was not 'in the possession and control of the Employer' since it was part of the highway.

7.11 Delay by nominated sub-contractors and statutory undertakers: JCT 80 clause 25.4.7

The inept words 'delay on the part of' are repeated in JCT 80 clause 25.4.7 in spite of numerous judicial criticisms and notwithstanding Lord Wilberforce's observation in *Westminster City Corporation* v. *Jarvis* (1970): 'I cannot believe that the professional body, realising how defective this clause is, will allow it to remain in its present form.'

The words do not mean delay *caused* by a nominated sub-contractor or supplier. Nor does delay mean just sloth or dilatoriness. Again, Lord Wilberforce said in *Westminster* v. *Jarvis*, 'it is contractually irrelevant whether a sub-contractor could have worked faster.'

It means solely and exclusively failure to complete the sub-contract works by the contractual date.

It does not include delay caused to the contractor by the repudiation or insolvency or defects of the sub-contractor.

As Lord Wilberforce said: 'If it were, why should the word "delay" be used? Why not frankly exonerate the contractor for any delay in completion due to any breach of contract or failure, *eo nomine* of the sub-contractor.'

In other words, if the draftsmen had intended that the contractor should get an extension of time for any delay *caused by the sub-contractor*, they would have used those words.

Above all, a contractor cannot rely on these words if the sub-contractor has ceased to work on the site. In *Westminster* v. *Jarvis* (*supra*) the piling sub-contractor had completed his work and withdrawn from the site when it was discovered that the piles were defective. 'The sub-contractor is not in delay so long as, by the sub-contract date, he achieves such apparent completion that the contractor is able to take over, notwithstanding that the work so apparently completed may in reality be defective.'

Once a sub-contractor has withdrawn from the site he cannot be 'in delay'.

The basic scheme of nomination, involving NSC/2 or 2a, seeks to give the employer who has lost his right to liquidated damages against the contractor recourse in damages against the nominated sub-contractor for breach of the warranty contained in NSC/2 clause 3.4 or NSC/2a clause 2.4. But that warranty must be read in the light of the provision of the contractor/sub-contractor contracts which in NSC/4 or 4a clause 11.2.2 requires the architect to consent to the contractor giving an extension of time to the nominated contractor for a whole string of relevant events set out in NSC/4 or 4a clause 11.2.5, plus the contractor's own default and in addition the valid exercise of the sub-contractor's right to suspend his work, contained in NSC/4 or 4a clause 21.8, for non-payment.

Two things are clear: if the employer wishes to recover lost liquidated damages he will have to sue the nominated sub-contractor and he will have extreme difficulties in proving a breach of the warranty.

The same attempt to exact a warranty from nominated suppliers is to be found in the Standard Form of Tender. Clause 25.4.7 qualifies delay on the part of the sub-contractor by the words 'which the contractor has taken all practicable steps to avoid or reduce'. It is the sub-contractor's delay, not his own, to which this subordinate clause applies.

For reasons set out in [4.16], the employer has no hope of recovering in contract lost liquidated damages from statutory undertakers where the contractor has secured an extension of time under JCT 80 clause 25.4.11 for:

> 'the carrying out by a local authority or statutory undertaker of work in pursuance of its statutory obligations in relation to the Works, or the failure to carry out such work.'

It might, however, be possible for the employer to frame an action in tort for negligence and/or breach of statutory duty.

7.12 Government action

A new ground for an extension of time is contained in JCT 80 clause 25.4.9:

> 'the exercise after the Date of Tender by the United Kingdom Government of any statutory power which directly affects the execution of the Works by restricting the availability or use of labour which is essential to the proper carrying out of the Works or preventing the Contractor from, or delaying the Contractor in, securing such goods or materials or such fuel or energy as are essential to the proper carrying out of the Works.'

7.13 Failure to obtain labour or goods

The provisions which were starred items in JCT 63 clause 23, and therefore optional, are now treated as mandatory. They include:

> 25.4.10.1 'the Contractor's inability for reasons beyond his control and which he could not reasonably have foreseen at the Date of Tender to secure such labour as is essential to the proper carrying out of the Works; or
>
> .2 the Contractor's ability for reasons beyond his control and which he could not reasonably have foreseen at the Date of Tender to secure such goods or materials as are essential to the proper carrying out of the Works.'

The perils of deleting these or other items in this clause have been pointed out already [3.04], as has the method of safely doing so. The date of tender would appear to be ten days before the date fixed for receipt of the tenders by the employer: see JCT 80 clause 1.3 which there refers solely to JCT 80 clauses 38.6.1 and 39.7.1. There is no definition specifically applicable to JCT 80 clause 25 and it may well be that there, what is meant is the actual date of tender. But by the definitions given in the Formula Rules it is ten days before the date fixed for the tenders. It may well be that for the purposes of JCT 80 clause 25 'the Date of Tender' is the ten days before but the draftsmen do not specify it in clause 1.3.

If one could apply the ordinary canons of construction to this contract, it would mean the opposite, on the basis that because JCT 80 clause 1.3 does not mention it; *expressio unius personae vel rei, est exclusio alterius* (that is, the express mention of one person or thing is the exclusion of another).

7.14 Events outside the control of the parties

Two items are specified in JCT 80 clauses 25.4.1 and .2.

Force majeure is a chameleon expression which takes its colour from the contractual background against which it is used. In this contract many of the events which are described as 'clause 22 perils' would in other contracts be regarded as *force majeure* and its meaning in this clause could well be different from that in JCT 80 clause 28.1.3.1. One thing is certain: if this expression is taken from this contract and transplanted into another (say, the JCT Minor Works Form which has no reference to 'clause 22 perils') another meaning entirely must be assigned to it.

Events which are outside the control of the contracting parties and outside their contemplation when they made the contract are normally *force majeure*: *Lebeaupin* v. *Crispin* (1920); *London Borough of Hackney* v. *Doré* (1922). But most events such as war, strikes and natural disasters of all kinds, which are normally *force majeure*, are specifically provided for in JCT 80. There appears to be no case in which the courts have been called upon to decide what it means in the context of this contract and cases about other contracts are no guide whatsoever to its meaning here.

The wording regarding the weather has been changed from 'any exceptionally inclement weather' to 'exceptionally adverse weather conditions' since it was apparently thought that the year of exceptional drought was not covered by the adjective 'inclement'.

Architects fail to give sufficient emphasis to the word 'exceptionally' and are prone to issue certificates far too readily when the weather, although it has impeded the contract works, is anything but exceptional for the area and period of year. It would be exceptional, for example, if there were not snow in North Yorkshire for most of January to March. Sometimes this generosity with the employer's money is due to sympathy with the contractor; at other times it appears to be a compromise to cover a refusal to issue a certificate under clause 25.4.6.

7.15 Loss or damage by clause 22 perils: JCT 80 clause 25.4.3

This being a contract prepared by the Joint Contracts Tribunal, 'Clause 22 perils' are to be found, not in clause 22, but in clause 1.3. They are: fire, lightning, explosion, storm, tempest, flood, bursting or overflowing of water tanks, apparatus or pipes, earthquake, aircraft or other aerial devices, or articles dropped therefrom, riot and civil commotion.

Earlier doubt was expressed whether, if some of these occurred by reason of the default of the contractor's own staff, he was entitled in spite of that to an extension of time. For example, if through their negligence there is a fire or explosion, the bursting or overflowing of water tanks etc. is the contractor still entitled to an extension of time?

It would appear so. JCT 80 clause 28 makes use of exactly the same words as JCT 80 clause 25.4.3 regarding determination but qualifies them by adding 'unless caused by the negligence of the contractor, his servants or agents or of any sub-contractor, his servants or agents'. Once again the normal presumption of the construction of contracts is that where words are used to qualify others in one place in a contract and not in another, then the omission is intentional: *expressio unius personae vel rei, est exclusio alterius*. But if the contractor burns down the works, the architect would be ill-advised to give him an extension of time, in spite of the indemnity in JCT 80 clause 20.2, since that is not wide enough to cover the loss of liquidated damages which are intended to compensate the employer for his loss of rent or loss of use of the premises. The plain tenor of the whole of JCT 80 clause 25 is that the contractor should not have an extension of time for anything that is in any way his fault.

7.16 Strikes and similar events: JCT 80 clause 25.4.4

'Civil commotion' has been defined as indicating a stage between a 'riot and a civil war' for insurance purposes. It is doubtful whether it is apt to cover, for example, interference with the work by nuclear protesters. It is to be noted that it is also a 'clause 22 peril' and the effect of this is considered in connection with the contractor's right to determine [9.08].

The wording regarded the effect of strikes is very wide, including 'any of the trades engaged in the preparation, manufacture or transportation of any of the goods or materials' required for the works but it is thought that it should be read as meaning a strike applicable to the whole of the trade concerned, and not to one particular firm in that trade.

7.17 Contractor's notice of delay

The procedure has apparently been tightened up by JCT 80 clause 25.1 in that:

(a) the contractor must forthwith give written notice if the progress of the work is likely to be delayed; and
(b) specify the cause or causes of the delay;
(c) if JCT 80 clause 25.4.7 is relied upon as a relevant event, notice has to be given to the nominated sub-contractor (but not apparently, strangely enough, to either a nominated supplier or to a statutory undertaker where JCT 80 clause 25.4.1 or 25.4.8 is relied upon); and
(d) the contractor must give particulars of the expected effects; and
(e) estimate the extent, if any, of the expected delay.

This clause certainly imposes a contractual obligation on the contractor and he will be in breach if he does not submit all the information.

It is still not clear whether compliance in all these matters is a condition precedent to granting an extension so that an architect would be entitled to refuse a notice that was not given 'forthwith' or which did not contain all the specified details.

The wording of JCT 80 clause 25.3.1 suggests that it is not, and that he has to consider any notice received whether in time or not and whether accompanied with the necessary information or not. It will be noted that a formal application for an extension of time is not necessary — a notice suffices.

The architect is required now to adjudicate upon notices within twelve weeks: JCT 80 clause 25.3.1 and there are severe consequences for his employer if he fails to do so since fluctuations continue. If the period between the receipt and the existing completion date is less than twelve weeks that completion date becomes the latest period for him to adjudicate. His obligation is to fix 'such later date as he then estimates to be fair and reasonable'. But this is qualified in that he shall do so 'if reasonably practicable having regard to the sufficiency of the notice, particulars and estimates'.

mean that the architect can just ignore such notices laim that he had insufficient information. He must em within twelve weeks by fixing a new date and of the relevant events he has taken into account, ent he has had regard to omissions: JCT 80 clause 25.3.1.3. Or, if he is unable to do that, by notifying the contractor that it is not reasonably practicable to fix a new date having regard to the lack of information, or that in his opinion none of the events relied upon is a relevant event, or that in his opinion the completion of the works is not likely to be delayed.

What he cannot now do is to ignore the notice, spike it and decide he will think about it later. The contract also provides that he must notify every nominated sub-contractor in writing of every decision about completion date; it is not limited to notification only of new dates.

7.18 Damages for non-completion: JCT 80 clause 24

The employer's right to deduct liquidated damages from monies otherwise due to the contractor is dependent upon the issue by the architect of a certificate to the effect that the contractor has failed to complete the works by completion date.

This is a condition precedent without which the employer has no right to make any deduction or recover any liquidated damages by action or otherwise.

The *JCT Guide* suggests (paragraph 15.1, page 12) that it is merely a 'confirmatory certificate of the completion date fixed after the architect has considered all applications for extensions of time under clause 25 and carried out the review in the twelve weeks after Practical Completion'.

The certificate of non-completion can and should be issued as soon as the effective completion date has passed — quite independent of and before the review required by JCT 80 clause 25.3.3.

The wording of JCT 80 clause 24.1 (though not JCT 63 clause 22) is mandatory on the architect:

> 'If the contractor fails to complete the Works by the Completion Date then the Architect *shall* issue a certificate to that effect' [author's italics].

This is made absolutely clear by JCT 80 clause 24.2.2:

> 'If, under clause 25.3.3 the Architect fixes a later Completion Date, the Employer shall pay or repay to the Contractor any amounts recovered allowed or paid under clause 24.2.1 for the period up to such later Completion Date.'

It has been held in the case of *Department of the Environment for Northern Ireland* v. *Farrans* (1981) that where the architect had issued a certificate under JCT 63 clause 22 and the employer had deducted liquidated damages, and the architect had subsequently granted a further extension of time, the contractor was entitled to interest on the monies deducted and repaid. But in this Northern Ireland case Mr Justice Murray held (a) that the effect was to make the deductions 'a breach of contract' by the employer and (b) that interest was recoverable as damages for that breach of contract.

So far as JCT 80 at any rate is concerned, that decision cannot be applicable because deduction after the architect's certificate is specifically authorised by JCT 80 clause 24.2.1. To say that if an employer does so, 'he does so at his own risk that a subsequent certificate . . . would vitiate the certificate on which it was relying and leave [him] without protection against a claim for breach of contract in failing to pay on the due date the amount certified,' as Mr Justice Murray did, can surely not be so even if applicable only to JCT 63. How can something that is expressly authorised by the contract ever be a breach of contract? JCT 80 clause 24.2.2 has a contractual provision as to what is to take place on the occurrence of specific events. The certificate is not vitiated; it is replaced under the provisions of the contract by another.

The new wording 'as a debt' appears to make it possible for the employer to set off these liquidated damages against any claim by the contractor arising out of this or any other contract; and this makes a material alteration as to what was provided by JCT 63 clause 22.

7.19 The architect's review: JCT 80 clause 25.3.3

The review of the completion date that the architect is required to carry out is described in JCT 80:

> 25.3.3 'Not later than the expiry of 12 weeks from the date of Practical Completion the Architect shall in writing to the Contractor either

.3.1 fix a Completion Date later than that previously fixed if in his opinion the fixing of such later Completion Date is fair and reasonable having regard to any of the Relevant Events, whether upon reviewing a previous decision or otherwise and whether or not the Relevant Event has been specifically notified by the Contractor under clause 25.2.1.1; or

.3.2 fix a Completion Date earlier than that previously fixed under clause 25 if in his opinion the fixing of such earlier Completion Date is fair and reasonable having regard to any instruction requiring as a Variation the omission of any work issued under clause 13.2 where such issue is after the last occasion on which the Architect fixed a new Completion Date; or

.3.3 confirm to the Contractor the Completion Date previously fixed.'

The architect has to carry out this review in the light of any relevant events whether or not they have been notified by the contractor. This will mean that in the case of clause 25.4.6, delay in the receipt of instructions etc., it is only a relevant event if there has been a specific application in writing on the due date. He has no discretion to deal with any other on his review.

7.20 Enforcing liquidated damages

'Payment of liquidated damages is no longer obligatory when the architect issues his certificate,' says the *JCT Guide*, paragraph 15.2, page 12. It never has been.

A note on page 22 with reference to JCT 80 clause 24 reads:

> 'This certificate merely confirms that the Completion Date has been passed without completion of the Works having been achieved: this may however be due to matters for which the Architect has no power under clause 25 (1980) to fix a new Completion Date but for which the Employer may be responsible; in such circumstances Employers may need to take advice as to the validity of any deduction of liquidated damages under clause 24.'

This cannot be so but is presumably an attempt to call employers' attention to the principles set out in this book in [7.02].

Finally paragraph 15.3, page 12 of the *JCT Guide* says: 'The Employer must give notice before the issue of the Final Certificate of his intention to exercise his discretion to claim or deduct liquidated damages.' No authority is quoted for this but presumably it is based upon the words in JCT 80 clause 24.2.1: 'as the Employer may require in writing not later than the date of the Final Certificate.' If so, it is surely a misreading of the clause which merely lays down that within that period the employer shall indicate whether he wants them paid or allowed.

The idea that liquidated damages are recoverable only if the employer gives notice that he requires them before the issue of the final certificate is incompatible with the fact of liquidated damages being a debt. They can clearly be deducted from the final certificate.

However, the clause is highly ambiguous and employers should claim them before final certificate to avoid argument.

Even suppose that liquidated damages were irrecoverable because they were not claimed before final certificate, this does not mean that the employer would forfeit his right to ordinary unliquidated damages for breach of contract by the contractor's failure to complete on time; and these might be, and usually are, much more than the fixed sum. But these provisions do not appear to contemplate that it may be possible for an employer not to require the liquidated damages to be paid, and then opt for much larger unliquidated damages for delay.

7.21 Variations after completion date

There is no doubt that under JCT 63, the architect could not order variations which would increase the work, or delay the contractor after the due date for completion had passed, without forfeiting the employer's right to liquidated damages. In *Bramall and Ogden Ltd* v. *Sheffield City Council* (1983) it was pleaded that the defendants could not recover liquidated damages because 'by issuing instructions involving variations and extra works after the date certified in the Clause 22 certificate and by failing to issue any extensions of time, the defendants . . . put time at large under the contract and thereby rendered null and void the . . . liquidated and ascertained damages' clause.

It was apparently conceded by the experienced leading counsel who appeared for Sheffield that, if that were indeed so, time would be at large.

The JCT Intermediate Form IFC 84 makes it clear that under that form the architect can order variations after the due date for completion. There can be little doubt that in JCT 80 the draftsmen thought that they had made similar provisions.

Regrettably, this is by no means certain.

In the first edition of this book the present author inclined to think that on the strict legal interpretation of the wording the view was correct that:

> 'because the contractor has not completed by the completion date, he is thereafter in breach of contract and if the architect issues a variation order or other instruction which further postpones completion, time becomes at large and the employer cannot deduct liquidated damages and is taken to have waived his rights to completion on the due date.'

The other argument is as follows.

Clause 13.2 gives the architect an unqualified power to order variations at any time.

Clause 25.4.5.1 gives him power to extend the date for completion for the contractor's compliance with clause 13.2 as a relevant event.

Clause 25.3.3 requires the architect not later than 12 weeks from the date of practical completion to fix a completion date *later* than that previously fixed, if fair and reasonable, having regard to any of the relevant events.

This can be done 'whether upon reviewing a previous decision . . . and whether or not the relevant event has been notified by the contractor'.

From this it is clear that the architect can put back the date for completion for variations ordered after the date for completion has passed and even if he has issued a certificate under clause 24.1.

This provides that if the contractor fails to complete the works by the completion date, he *shall* issue a certificate to that effect and thereafter the employer may deduct liquidated damages. This, of course, is the equivalent to the certificate under JCT 63 clause 22.

But there is one vital difference. Clause 24.2.2 provides:

> 'If, under clause 25.3.3, the Architect fixes a later Completion Date, the Employer shall pay or repay to the Contractor any amounts recovered allowed or paid under clause 24.2.1 for the period up to such later Completion Date.'

It follows that, by this tortuous method, it can be concluded that the architect can issue variation orders even after he has issued a certificate under clause 24.1 and still preserve his employer's rights to liquidated damages from a later completion date. It is a great pity the Tribunal could not have said so directly.

It now appears that under JCT 63 the architect could not issue variation orders after the due completion date without forfeiting the employer's right to liquidated damages; but under JCT 80 he can, provided he adjusts the completion date.

Chapter 8

Insurance provisions

8.01 Varying the printed contractual terms

A major problem regarding the insurance provisions of JCT 80 arises out of JCT 80 clause 2.2.1:

> 'Nothing contained in the Contract Bills shall override or modify the application or interpretation of that which is contained in the Articles of Agreement, the Conditions or the Appendix.'

For some reason, the wording has been slightly altered from that which appeared as JCT 63 clause 12(1). There the words were 'nothing contained in the Contract Bills shall override, modify or affect in any way whatsoever . . . these Conditions'. Although less emphatic, it is thought that there has been no material alteration in meaning.

It may seem strange to discuss this in connection with the insurance provisions of the contract, but it is in this connection that most cases dealing with JCT 63 clause 12(1) have arisen. As will be seen, the insurance provisions are manifestly unsatisfactory and it is common practice to make provision in the bills of quantities to include more satisfactory provisions. As the second edition of this book goes to press, they are under review by the JCT.

Normally, written or typewritten documents specially prepared for a contract and used in conjunction with it take precedence over printed forms: see Scrutton LJ in *Sutro & Co* v. *Heilbut Symons & Co*. (1917) and *Love & Stewart Ltd* v. *Rowtor SS Co. Ltd* (1916). The typed bills therefore should prevail over the printed Articles and Conditions.

In *Gold* v. *Patman & Fotheringham Ltd* (1958) Lord Hodson said of a similarly worded clause 10 of the 1939 RIBA contract:

> 'The paragraph [in the bills of quantities] beginning with the words "such insurance" purports to override the provisions as to insurance contained in Condition 15 and must therefore be disregarded, having regard to Condition 10.'

Again, in *North West Metropolitan Hospital Board* v. *Bickerton* (1970) the same judge said:

> 'I have not referred to clause 16(f) in the bills of quantities . . . in view of the difficulties presented by Condition 12(1) [of JCT 63] which makes it clear . . . that the Conditions cannot be overridden, modified or affected by what is contained in the bills.'

Judge Stabb QC, the senior Official Referee, took the same view in *Mowlem* v. *BICC Pension Trust* (1977), as did Mr Justice Mocatta in *Gleeson* v. *Hillingdon* (1970).

The same point was at issue in the Court of Appeal in *English Industrial Estates* v. *George Wimpey & Co. Ltd* (1973). The defendents entered into a JCT 63 contract to extend a factory in Hartlepool that was owned by the plaintiffs and occupied by their tenants, Reed Corrugated Cases Ltd. Clause 20A(1) required Wimpey's to insure against fire and other risks. It also provided in clause 16 that if the employer took possession of any part of the works before practical completion, the contractor should reduce the value insured by 'the full value of the relevant part' and the said relevant part 'should be at the sole risk of the Employer'. Clause 18 of JCT 80 is in the same terms.

A fire took place in January 1970 which destroyed various buildings which were nearly completed but which were already in use by the tenants, with the contractor's consent.

The bills of quantity in fact contained provisions that the contractor would allow the employer or anybody authorised by him 'as much plant and equipment' in buildings as was possible before the completion of the work.

Further it provided, again in the bills, that if because of the installation of such plant and equipment, the cost of insurance to the contractor was increased, the increased cost should be added to the contract sum.

English Industrial Estates argued that because of that provision in the bills of quantities the loss fell on the contractors. Wimpey's argued that because of clause 12(1) the provisions in the bills should be ignored, as modifying the printed contract conditions and by virtue of clause 16, the employer had taken possession of the premises and they were at the risk of English Industrial Estates.

Lord Denning took the view that type should *always* prevail over print and therefore the provisions of the bills of quantity prevailed over clause 12(1) and clause 16 of JCT 63.

Lord Justice Edmund-Davies, as he then was, pointed out that the bills were in this case what he termed 'a curiously hybrid document' and that no less than six clauses of JCT 63 were amended by a part of the bills called 'The Form of Contract'. Of the insurance clauses quoted above he said: 'It is difficult to see why these provisions appear at all unless they were intended by the parties to have some contractual efficacy.'

He adopted a compromise position: 'For my part, I cannot accept that *some* regard may not be had [to the Bills].'

They might be used, he held, 'not in the interpretation of the contract (a purpose for which they may not be utilised) but in order to follow what exactly was going on'. He therefore concluded that use by the tenant was not taking possession by the employer.

It was in this judgment that his lordship referred to 'the farrago of obscurities which make up' the JCT 63 contract.

Lord Justice Stephenson expressed the hope that those responsible for the contract would redraft JCT 63 clause 16 to bring 'greater clarity, consistency and certainty into the contract'. This hope has not been achieved since, as the *JCT Guide* reports at page 21: 'Clause 16 (1963) is substantially reproduced as clause 18 (1980)' except for the fact that 16(f) regarding release of retention has been omitted.

His lordship also called for a revision of JCT 63 clause 12(1), a plea which has also been ignored. As to the provisions of that clause, he said:

> 'I confess with diffidence that I cannot go all the way with the Master of the Rolls. To apply the general principle that type should prevail over print seems to me to contradict the express provisions of clause 12(1) that the reverse is to be true of this particular contract: the special conditions in type are to give way to the general conditions in print.'

He thought that words in the bills of quantities:

> 'in so far as they state conditions of the contract, have no effect on the printed conditions.

> In so far as they introduce further contractual obligations . . . they may add obligations which are consistent with the obligations imposed by the conditions, but they do not affect them by overriding or modifying them or in any other way whatsoever.
>
> It follows that . . . the court must disregard — or even reverse — the ordinary and sensible rules of construction. . . .'

In that case, therefore, three judges in the Court of Appeal arrived at the same conclusion, namely that the contractors were liable for the loss sustained by fire, for buildings occupied by the employer's tenant, but all arrived at that conclusion by a different route.

Subsequent cases suggest that, notwithstanding Lord Denning's contentions, the position in law now is:

(1) JCT 80 clause 2.2 must be given full effect in that the printed conditions prevail over the typed documents, so that the normal rule is reversed.
(2) Typed documents, such as the bills of quantities, cannot be used for the interpretation, still less the modification or alteration, of the printed contract, but they can be looked at to see 'what exactly was going on'.

This approach was adopted by Judge Fay, sitting as a deputy Queen's Bench judge in *Henry Boot* v. *Central Lancashire New Town Development Corporation* (1981), when he was considering whether work to be executed by statutory undertakers and not in performance of their statutory duties was work 'not forming part of this contract' [7.09].

How then can an employer vary the terms of JCT 80? Not, clearly, by writing in terms in the specification or bills of quantities.

Lord Justice Stephenson indicated the only way:

> 'Those building owners who put forward the RIBA contract ought . . . to see either that clause 12 is struck out or amended or that all contractual terms which should be read together with the printed conditions are not left in the bills of quantities only but are incorporated in the Conditions . . . Otherwise, building owners have only themselves to blame.'

To which may be added the caution that in those circumstances the employer may justly blame — and sue — his specialist advisers.

8.02 Contractor's indemnity to employer for personal injuries and death

JCT 80 clause 20 provides an indemnity by the contractor to the employer in respect of claims arising from personal injuries to anybody or the death of anybody 'arising out of or in the course of or caused by the carrying out of the Works'.

In spite of the fact that the contractor is normally in possession of the site, that possession may not be exclusive of the employer's possession and the employer may therefore be sued under the Occupiers Liability Act 1957 or at common law or vicariously, as being responsible for the acts or omissions of the contractor, even though he is an independent contractor and not a servant.

An indemnity is, of course, only operative after the employer has been condemned by a judgment against him. But the employer can take out third party proceedings against the contractor before judgment is actually given against him: *County & District Properties* v. *Jenner* (1974).

From this indemnity, the clause exempts injuries or death 'due to any act or neglect of the Employer'.

That is fair enough and fair enough that it should extend to acts of negligence by the employer's own servants for whom he is vicariously liable. But it also exonerates the contractor in respect of persons who are not the employer's servants but those 'for whom the employer is responsible'.

These may fall into two categories: those for whom the employer is to be deemed to be responsible under the express terms of the contract and others for whom the employer is in fact and in law vicariously responsible.

Those falling within the first category by the terms of JCT 80 clause 29.3 are as follows:

> 'Every person employed or otherwise engaged by the Employer . . . shall for purpose of clause 20 be deemed to be a person for whom the Employer is responsible. . . .'

As has been seen, if the judgment in *Henry Boot* v. *Central Lancashire New Town Development Corporation* (1981) is correct this can include statutory undertakers doing work outside their statutory obligations.

Are the architect, quantity surveyor and the clerk of works persons 'for whom the employer is responsible'? The question has been raised and so far there is no satisfactory answer.

8.03 Liability under indemnities in general

Is the indemnity effective to cover injuries which result in part from the conduct of the employer or his servants or agents?

There is authority that these words do not exempt the contractor for any 'act or neglect' on the part of the employer etc. unless that act or neglect amounts to negligence at common law: *Hosking* v. *De Havilland Ltd* (1949). A claim against the employer for breach of statutory duty has to be indemnified by the contractor: *Murfin* v. *United Steel* (1957).

The indemnity in this clause may well be valueless to the employer for it is an established principle of indemnity law that unless an indemnity clause expressly covers the negligence of the party being indemnified, the presumption is that it is not intended to saddle the one giving the indemnity with responsibility for the indemnified's own negligence which in part contributed to the loss: *Alderslade* v. *Hendon Laundry* (1945); *Smith* v. *South Wales Switchgear* (1978). This principle has been applied to construction industry cases in *Walters* v. *Whessoe Ltd and Shell* (1960), an unreported case to which reference was made in *AMF (International)* v. *Magnet Bowling and Ano.* (1968) which dealt with clause 14(b) in the 1957 RIBA contract, which was in similar terms to JCT 80 clause 20.

There it was held that because the contractor and the building owner were found liable as joint tortfeasors under the Occupiers Liability Act 1957 to the plaintiff, assessed under the Law Reform Act (1935) as 40 per cent employer and 60 per cent contractor, the indemnity clause was of no effect. The force of this decision, however, is somewhat vitiated by the fact that Mr Justice Mocatta, although he had correctly formulated the duty of the employer's architect, seems to have misapplied the test to the facts and came to a conclusion as to the employer's liability which his own argument did not justify. It has been suggested that this decision is not consistent with the House of Lords' decision in *White* v. *Tarmac Civil Engineering* (1967) in which plant hirers were held entirely liable under the contract, even though the injury was in part the fault of the plant owner. However, the contrary is in fact the case since counsel did not rely upon the indemnity clause in view of the *Walters* v. *Whessoe Ltd & Shell* case *supra*, which was regarded as authoritative.

The position appears to be that all indemnity clauses are to be construed strictly as exemption clauses *contra proferentem* the party in whose favour the indemnity is granted: *AMF* v. *Magnet Bowling Ltd and Trentham Ltd* (1968); and *Manchester* v. *Fram Gerard Ltd* (1974).

If the indemnity clause purports to hold a party liable for defaults etc. other than his own, the person indemnified cannot rely upon it unless it spells this out so as to make it quite clear that the one giving the indemnity is to be responsible for those over whom he has no control: *Gillespie Brothers* v. *Roy Bowles Transport* (1973).

The position may be further complicated by the Civil Liability (Contribution) Act 1978 which provides in section 7(3):

> 'The right to recover contributions in accordance with section 1 above supersedes any right, other than an express contractual right, to recover contributions (as distinct from indemnity) otherwise than under this Act in corresponding circumstances; but nothing in this Act shall affect
> (a) any express or implied contractual or other right to indemnity; or
> (b) any express contractual provision regulating or excluding contributions;
> which would be enforceable apart from this Act (or render enforceable any agreement for indemnity which would not be enforceable apart from this Act).'

If it is thought that the indemnity in JCT 80 clause 20 may be unenforceable for the reasons given, this section does not in any way alter the courts' power to assess such contribution between the parties 'as may be found by the court to be just and equitable having regard to that person's responsibility for the damage in question', including exempting them from all liability or directing a complete indemnity': section 2, Civil Liability (Contribution) Act 1978.

In these circumstances it is difficult to see what value in any circumstances JCT 80 clause 20 is to the employer, except perhaps to extend the contractor's possible period of contractual liability under the Limitation Act 1980.

A similar indemnity is that contained in JCT 80 clause 20 given by each nominated sub-contractor to the contractor in respect of sub-contract works in NSC/4 or 4a clause 6; the same considerations apply to it.

8.04 Contractor's indemnity in respect of injury to property

Clause 20.2 provides that:

> '. . . the Contractor shall be liable for, and shall indemnify the Employer against, any expense, liability, loss, claim or proceedings in respect of *any injury or damage whatsoever to any property real or personal* in so far as such injury or damage arises out of or in the course of or by reason of the carrying out of the Works . . .' [author's italics].

There are two qualifications to those sweeping words in the clause itself. The first is the proviso:

> 'Except for such loss or damage as is *at the sole risk* of the Employer under clause 22B or 22C . . .' [author's italics].

The second is the other proviso:

> '. . . provided always that the same is due to any negligence, omission or default of the Contractor, his servants or agents, or of any sub-contractor, his servants or agents.'

Therefore fault of some kind by the contractor or a sub-contractor is an essential ingredient. It would also appear that the clause may well cover sub-sub-contractors: *Petrofina (UK) Ltd* v. *Magnaload Ltd* (1983). That case was primarily concerned with the wording of a contractor's so-called 'all risks' policy issued by the New Hampshire Insurance Company in which the insurers were attempting to secure an indemnity by way of subrogation from sub-sub-contractors who, the courts held, were in fact insured under the policy, which covered 'contractors and/or sub-contractors'.

There is also a further qualification established by the courts. The words are not wide enough to cover the contract works themselves: *Tozer Kemsley Millbourn (Holdings) Ltd* v. *Cavendish Land (Metropolitan) Ltd and General Insurance Society Ltd* (1983).

Questions which arise under this clause are whether it covers trespass to adjacent land or air space, e.g. by an overhead crane; removal of a right of support; nuisance to adjacent land; or escape of materials, as in *Rylands* v. *Fletcher* (1868).

Trespass to land is actionable *per se*, without proof of damage or fault. Trespass to goods requires fault on the part of the tortfeasor: *National Coal Board* v. *Evans* (1951).

Nuisance was at one time thought not to be actionable unless knowledge or negligence was proved. It now appears to be accepted that neither is necessary before an action will lie: *Dodd* v. *Canterbury City Council* (1980); *Lord Advocate* v. *Reo Starkis Organisation Ltd* (1981), except in the case of things naturally on the land.

In the latter case, piling operations were carried out by a sub-contractor at 192–201 Ingram Street, Glasgow, which was separated by a lane from Lanarkshire House. It was claimed that the piling operations had done damage to Lanarkshire House by disturbing the natural support. It was held that an action would lie for nuisance done by building operations, just as it would for any other type of use which resulted in injury to adjacent or adjoining property, and that an action would lie without proof of knowledge, intent, negligence or foreseeability. Although this was a Scottish case, it was said that the law in England was to like effect. It was, it will be noted, the owners, employers of the main contractor, who were held liable.

There can be no doubt that the *Rylands* v. *Fletcher* liability is a strict liability, regardless of negligence or fault.

The question therefore is whether, under this clause, the contractor indemnifies the employer if the employer is held liable for any of these torts. It all depends upon the word 'default' in this context.

The word was defined by Mr Justice Parker in *re Bayley-Worthington and Cohen's Contract* (1909):

> 'Default must involve either not doing what you ought or doing what you ought not to do, having regard to your relations with other parties concerned in the transaction; in other words, it involves the breach of some duty you owe to another or others. It refers to personal conduct and is not the same thing as breach of contract.'

Mr Justice Kerr in *Manchester* v. *Fram Gerard* (*supra*) adopted this and added:

> 'Default would be established if one person covered by the clause either did not do what he ought to have done or did what he ought

> not to have done in the circumstances and that the conduct in question involves something in the nature of a breach of duty.'

It is clear therefore that while breach of contract and breach of a duty to take care are comprehended in this word, as also are breaches of statutory duty such as those contained in the building regulations and the Health and Safety at Work Act, the word is wider than that and includes breaches of duties which are not actionable in contract or in tort.

Since the only indemnity given to the employer under clause 20.2 is that in respect of 'negligence, omission or default of the contractor . . . or any sub-contractor', it would appear that the employer is without protection in respect of the very things for which he needs protection, i.e. torts such as nuisance whereby damage is caused to adjoining or adjacent property by reason of work on his own land, where liability exists independently of fault.

8.05 Insurance requirements of JCT 63 clause 19(2)(a); now JCT 80 clause 21.2.1

The name of this clause has become so well-known in insurance circles that it is retained here for convenience. It is now JCT 80 clause 21.2.1. The original JCT 63 clause was amended in 1968, and JCT 80 clause 21.2.1 has also been amended since it was first published, but without formal notice.

It now reads:

> 21.2.1 'Where a provisional sum is included in the Contract Bills in respect of the insurance to which clause 21.2 refers the Contractor shall maintain in the joint names of the Employer and the Contractor insurances *for such amounts of indemnity* as may be specified in the Contract Bills in respect of any expense, liability, loss, claim or proceedings which the Employer may incur or sustain by reason of damage to any property other than the Works caused by collapse, subsidence, vibration, weakening or removal of support or lowering of ground water arising out of or in the course of or by reason of the carrying out of the Works *excepting damage*:
>
> .1.1 caused by the negligence, omission or default of the Contractor, his servants or agents or of any sub-contractor, his servants or agents;

.1.2 attributable to errors or omissions in the designing of the Works;
.1.3 which can reasonably be foreseen to be inevitable having regard to the nature of the work to be executed or the manner of its execution;
.1.4 which is at the sole risk of the Employer under clause 22B or 22C (if applicable);
.1.5 arising from a nuclear risk or a war risk or sonic booms.
21.2.2 Any such insurance as is referred to in clause 21.2.1 shall be placed with insurers to be approved by the Employer, and the Contractor shall deposit with him the policy or policies and the receipts in respect of premiums paid.' [author's italics.]

The clause originated as the result of the case of *Gold* v. *Patman and Fotheringham* (1958), which was concerned with the 1939 RIBA Form (1952 revision).

Clause 14 required the contractor to give indemnities to the employer similar to those now contained in JCT 80 clause 20.

Clause 15 required the contractor to 'effect . . . such insurances . . . as may specifically be required by the bills of quantities'.

The bills of quantities required 'insurance of adjoining properties against subsidence or collapse'.

Piling sub-contractors, without negligence, removed support from adjacent land and the employer was held liable in an action for nuisance. It then appeared that the contractors had taken out a policy which covered them and their sub-contractors for causing subsidence to adjacent property but which did not cover the employer.

The employer therefore sued the contractors for breach of clause 15.

The Court of Appeal, in common with the trial judge, held that 'that which was insured against was the legal liability to adjoining owners arising out of the subsidence or collapse of the [adjacent] premises'.

'The only real difficulty which arises,' said Lord Justice Hodson, 'is as to the determination of the question as to who is to be covered by that insurance.'

He held that the contract specified what insurances had to be in the joint names of the employer and the contractor, but that the bills of quantity contained no obligation for the contractors to cover anybody but themselves.

It had been argued that, by reason of the indemnity clauses, there was an implied obligation that the insurance of the adjacent premises should cover both the employer and the contractor; this argument was rejected.

The employer in fact pays out the premium under JCT 63 19(2)(a) for nothing, for the simple reason that the policy is not required to cover, by clause 21.2.1.1, the only circumstances in which the contractor is liable to indemnify the employer under clause 20.

As has been seen [8.04], the contractor gives an indemnity to the employer only in respect of damage caused to other property due to any negligence, omission or default of the contractor, his servants or agents or of any sub-contractor, his servants or agents: clause 20.2. He is required by clause 21.2.1 to insure in joint names only in respect of 'such amounts of indemnity' which are *not* caused by 'the negligence, omission or default of the contractor, his servants or agents or of any sub-contractor, his servants or agents' etc. He does not give an indemnity in respect of this.

However, there is great scope for argument as to damage 'which can reasonably be foreseen to be inevitable having regard to the nature of the work to be executed or the manner of its execution'. In one sense, it can be said that all excavations and all pile driving inevitably create the risk of *some* damage, however trifling, to adjacent or adjoining properties; even the removal of an existing building can lead to an alteration in water levels in the ground, with consequent liability to an adjacent building.

The clause was drafted in 1968 by the British Insurance Association (BIA). The *JCT Guide* page 22, states that the BIA has said:

> 'The necessary cover is often available by extensions of the Contractor's Public Liability policy rather than the issue of separate policies. Cover granted in this way is always in the exact terminology of the present clause JCT 63 19(2)(a) and therefore complies as fully with the insurance requirements as will be the case if a separate policy is effected.'

As will be seen, there are so many exemptions in this clause that it is not surprising that claims under it are virtually unknown. Since under the clause the employer has to pay the premiums anyway, it is suggested that he would be much better served by taking out his own cover against actions likely to be brought against him for nuisance, trespass, withdrawal of support, *Rylands* v. *Fletcher* liability, and any other possible liability of this nature, however caused.

8.06 Insurance against injury to persons and property: JCT 80 clause 21

In addition to his indemnity to the employer, the contractor is also required to carry insurance against injuries of this nature: JCT 80 clause

21.1.1.1 'Without prejudice to his liability to indemnify the Employer under clause 20 the Contractor shall maintain and shall cause any sub-contractor to maintain such insurances as are necessary to cover the liability of the Contractor or, as the case may be, of such sub-contractor in respect of personal injury or death arising out of or in the course of or caused by the carrying out of the Works not due to any act or neglect of the Employer or of any person for whom the Employer is responsible and in respect of injury or damage to property, real or personal, arising out of or in the course of or by reason of the carrying out of the Works and caused by any negligence, omission or default of the Contractor, his servants or agents or, as the case may be, of such sub-contractor, his servants or agents.

.1.2 The insurance in respect of claims for personal injury to, or the death of, any person under a contract of service or apprenticeship with the Contractor or the sub-contractor as the case may be, and arising out of and in the course of such person's employment, shall comply with the Employer's Liability (Compulsory Insurance) Act 1969 and any statutory orders made thereunder or any amendment or re-enactment thereof. For all other claims to which clause 21.1 applies the insurance cover shall be the sum stated in the Appendix (or such greater sum as the Contractor may choose) for any one occurrence or series of occurrences arising out of one event.

21.1.2 As and when he is reasonably required so to do by the Employer the Contractor shall produce and shall cause any sub-contractor to produce for inspection by the Employer documentary evidence that the insurances required by clause 21.1 are properly maintained, but on any occasion the Employer may (but not unreasonably or vexatiously) require to have produced for his inspection the policy or policies and premium receipts in question.

21.1.3 Should the Contractor or any sub-contractor make default in insuring or in continuing or in causing to insure as provided in clause 21.1 the Employer may himself insure against any risk with respect to which the default shall have occurred and may deduct a sum or sums equivalent to the amount paid or payable in respect of premiums from any monies due to or become due to the Contractor or such amount shall be recoverable from the Contractor by the Employer as a debt.'

The requirements of the contract do not specify for how long such insurance should be maintained. The contractor should note that liability to indemnify the employer may extend to long after the contractor has completed the contract, since under the Limitation Act 1980 the cause of action only accrues when the employer has suffered loss, i.e. by a judgment against him [2.13].

8.07 The alternative insurance provisions

The JCT 80 contract provides, as did its predecessor, for three alternative methods of insurance of the works and premises themselves.

JCT 80 clause 22A is intended for adoption where it is the contractor who is to insure in the joint names of himself and the employer the works on a new build site.

Clause 22B is likewise to be used for new build works when it is intended that they shall be at the sole risk of the employer. This is not a contractual requirement in the local authority edition, but is in the private edition of JCT 80.

Clause 22C is for renovation work, or extensions or alterations to existing buildings. The risk is then placed on the employer.

The situation regarding consequential loss is not covered in any of these contractual terms.

Professional fees for reinstatement are mentioned but do not cover fees necessary for the presentation of claims.

8.08 Materials and goods: an insurable interest?

JCT 80 clause 22A.1 requires the contractor in joint names to insure not only all work executed and '*all* unfixed material and goods delivered to, placed on or adjacent to the works and intended for incorporation therein' (author's italics).

Similarly in clauses 22B.1 and 22C.1 cover by the employer is required for '*all* unfixed materials and goods delivered to, placed on, etc.' (author's italics).

Attention must therefore be drawn to the fact that in the case of JCT 80 clause 22A neither the contractor nor the employer may have an insurable interest in such materials; and in the case of JCT 80 clauses 22B.1 and 22C.1, the employer may not have an insurable interest in such materials.

The result may be that premiums are to be paid under JCT 80 in respect of materials, claims for which would not be recoverable in the event of a loss. Risk normally passes with the ownership of the goods (section 20, Sale of Goods Act 1979) but risk does usually pass to a buyer once the goods are delivered, even if payment is delayed: *Comptoir d'Achat et de Vente* v. *De Ridder* (1949).

In the Queensland case of *Knoblauch* v. *McInnes* (1935) stoves and gas coppers supplied to a builder were held to be at his risk as purchaser even though the contract provided for them to be paid for only when the houses in which they were to be installed were sold.

A well drafted retention of title clause will, of course, pass the risk to the purchaser even if the property is not to pass until payment. In those circumstances, the employer or the contractor may have an insurable interest in goods supplied. That is the position regarding sale of goods. The difficulty is that, under the JCT contracts, neither the contractor nor the employer is the buyer of goods brought on site by a work and materials sub-contractor. There might possibly be liability for unfixed materials as bailees, and in those circumstances an insurable interest, but it will be an issue of fact whether or not the contractor or the employer is ever the bailee of a sub-contractor's materials. In most circumstances, it is submitted, the contractor is not, and the employer never is, since neither has possession of the goods of a sub-contractor as opposed to possession of the site.

The position is further complicated by the provisions of the standard sub-contracts NSC/4 or 4a:

8.1.1.1 'Where clause 22A of the Main Contract Conditions applies the Contractor shall effect the insurance required by clause 22A, namely to insure against loss and damage by the "Clause 22 Perils" all work executed and all unfixed materials and goods delivered to, placed on or adjacent to the Works and intended for incorporation therein but excluding temporary buildings, plant, tools and equipment owned or hired by the contractor or any sub-contractor and keep such work, materials and goods so insured, etc.

.1.2 The sub-contractor shall not be responsible for any loss or damage however caused where clause 22A of the Main Contract Conditions applies or be under any obligation to effect any insurance in respect of clause 22 perils.'

In other words, the contractor is required to insure the sub-contractor's materials and goods in which he has no insurable interest whatsoever. The only way in which he can be said to acquire an insurable interest is under NSC/4 or 4a:

8.2.1 'Where clause 8.1.1 applies, then in the event of any loss or damage being caused by any one or more of the "Clause 22 Perils" to the Sub-Contract Works and any of the materials or goods of the Sub-Contractor delivered to, placed on or adjacent to the Sub-Contract Works and intended for incorporation therein the Contractor to the extent of such loss or damage shall pay to the Sub-Contractor the full value of the same, such values in the case of loss or damage to the Sub-Contract Works to be calculated as if the reinstatement of the loss or damage had been carried out in accordance with instructions of the Architect as to the expenditure of a provisional sum.'

In other words, because the contractor has had to insure them and when he has done so, he has to pay the sub-contractor the full value of 'the materials'. This does not constitute an insurable interest which is one that must exist at the time of the placing of the policy. This obligation to pay the sub-contractor arises only when there has been loss under the policy. Can a man insure another's goods, in which he has no insurable interest, just because he promises to pay over to the other the proceeds of any loss? Most certainly not.

It is also difficult to reconcile these provisions with NSC/4 or 4a:

8.3 'Subject to clause 8.1

8.3.1 The Sub-contractor shall be responsible for loss or damage to all materials or goods properly on site for incorporation in the Sub-Contract Works (except any such materials and goods as have been fully, finally and properly incorporated into the Main Contract Works) other than for any loss or damage due to any negligence, omission or default of the Contractor, his servants or agents, or any other sub-contractor of the Contractor engaged upon the Main Contract Works or any part thereof, his servants or agents or of the Employer or any person for whom the Employer is responsible.

.2 Where materials or goods have been fully, finally and properly incorporated into the Main Contract Works before practical completion of the Sub-Contract Works the Contractor will be responsible for loss or damage to such materials and goods except for any loss or damage caused thereto by the Sub-Contractor, his servants or agents.

.3 On practical completion of the Sub-Contract Works the Contractor will be responsible for loss or damage to the Sub-Contract Works properly completed and handed over except for any loss or damage caused thereto by the Sub-Contractor, his servants or agents.'

The idea is quite clear: the contractor is to insure the sub-contractor's goods and materials, even though he does not own them and never will, against clause 22 perils only. All other risks, such as theft and vandalism, are to be the sub-contractor's until the work has been incorporated in the realty.

Difficulty has arisen about the meaning of the words 'fully, finally and properly incorporated', but it appears that this is fulfilled by attachment to the realty in the place intended, even if they subsequently have to be moved.

The situation becomes even more bizarre when by NSC/4 and 4a clause 8.1.2.1 where JCT 80 clause 22B or 22C is in use, neither the contractor nor the sub-contractor is required to insure the sub-contractor's unfixed materials against clause 22 risks. Presumably, they are to rely upon the employer's insurance of materials in which the employer has no insurable interest. Where such loss occurs, the replacement is to be treated as a variation required by the architect under JCT 80 clause 13.2.

None of this can in any way be binding on the employer since he is not a party to NSC/4 or 4a and the contract expressly says in clause 5.2:

'Nothing contained in the Sub-Contract Documents shall be construed as to impose any liability on the sub-contractor in respect of any act or omission or default on the part of the employer . . . nor . . . create any privity of contract between the sub-contractor and the employer. . . .'

The sub-contractor/employer sub-contracts NSC/2 or 2a do not mention insurance.

The net result of all this is that contractors and employers may be paying out premiums for unsecured materials owned by sub-contractors which are in fact uninsured against clause 22 perils.

8.09 Future revisions to insurance provisions

At the time of writing the Joint Contracts Tribunal, not before time, has under consideration a complete revision of the insurance provisions. Particular care should therefore be taken when reading this book to ensure that the provisions of any particular contract are in accordance with those set out in JCT 80 as originally published, and not some subsequent version.

Chapter 9

...ermination before completion

9.01 Discharge before completion by breach of contract

Not every breach of contract by one party will entitle the other to refuse to perform his own obligations. Before there can be discharge by reason of the other party's breach, that breach must be such as evinces an intention to repudiate the whole of the contractual obligations. Such an intention may be indicated by express words or by 'repudiatory conduct' which effectively prevents the other party performing what he has promised to do. For example, this happens if the employer refuses or delays to hand over the site to the contractor: *Roberts* v. *Bury Commissioners* (1870); *Freeman* v. *Hensler* (1900); *Smart & Co*. v. *Rhodesia Machine Tools* (1950). If he failed to produce drawings, this would inevitably be held to be *indicia* of a breach of contract which would discharge the contractor from further performance.

When there is an express anticipatory repudiation before the time for performance comes round — 'I refuse to go on with the contract' — the innocent party generally has the option of accepting or rejecting the repudiation. 'An unaccepted repudiation is writ on water' is how one judge described it. The innocent party can either accept this repudiation and sue at once for damages; or he can continue to perform his obligations under the contract and hold the other party liable for all the money due under the contract: *White & Carter* v. *McGregor* (1961).

This, of course, only applies where the innocent party is capable of performing his own obligations under the contract. In the case of employees wrongfully dismissed, they cannot possibly go on working for their employer, and in the case of contractors who are refused possession of the site, clearly they have no option but to accept the repudiation because, in addition to the contractual right, they also need a property one — a licence to enter upon and occupy the site.

The position of contractors who are in possession of the site and who could continue to operate, in spite of wrongful express repudiation by the employer, is more difficult. In theory, they are entitled to refuse to accept the repudiation and continue in possession of the site and working to the plans provided. Mr Justice Megarry, as he then was, in *London Borough of Hounslow* v. *Twickenham Garden Developments Ltd* (1970) refused to grant an interlocutory injunction to give the council possession of the site. He decided the case on the basis of property law rather than contractual rights. His decision attracted a lot of criticism, including the observation in 87 Law Quarterly Review 309:

> 'The contractor's sole object is to make a profit; if he gets in, and suffers no loss, why should not the owner be entitled to change his mind and exclude him from the property?'

That was a very ill-founded observation. A contractor's *sole* object is not to make a profit. He has his reputation to consider which is a great deal more than making a profit.

With respect to his Lordship, the present author, who at the time of the decision was disposed to agree with the criticisms, now considers that Mr Justice Megarry was undoubtedly right.

Mr Justice Megarry held as he did for two reasons. Firstly, the contract for specified works over a specified period continued and there was an implied obligation on the part of the council not to revoke the contractor's licence, otherwise than in accordance with the contract. A Court of Equity, by granting an interlocutory injunction, would be assisting the council to break its contract. Secondly, although there was some case for saying that the council might have validly terminated the employment of the contractor under the terms of the contract, that was by no means self-evident and the *status quo* should be preserved until that issue was resolved. Mr Justice Megarry said:

> 'The case falls considerably short of any standard upon which . . . it would be safe to grant this injunction on motion . . . with so much turning on disputed questions of facts and inferences from facts, I cannot say that the borough has made out its case for the injunction that it seeks.'

An injunction to remove the contractors from site would, far from preserving the *status quo*, have prejudiced the contractor's case in a matter in which it was far from clear whether the employer had any right to determine the contract; and for that, damages to the contractor would have been an inadequate remedy.

The New Zealand case of *Mayfield Holdings Ltd* v. *Moana Reef Ltd* (1973), which refused to follow this decision, produced no convincing reasons why this view was wrong, particularly since the trial judge said, 'I am not prepared to hold . . . that there was a covenant implied in the contract binding the owner not to revoke the contractor's licence in breach of contract.' It must be self-evident that there is such an implied covenant. No Court of Equity should assist an employer in what is manifestly a breach of contract.

Non-payment by employer

Non-payment by the employer of sums certified by the architect, even though it constitutes a breach of contract, is rarely taken to evince an intention to repudiate.

The law regarding progress payments and payment by instalments was laid down by the House of Lords in the case of *Mersey Steel and Iron Co.* v. *Naylor Benzon & Co.* (1884). Failure to pay for one or more previous deliveries did not exonerate the other party from the obligation to deliver subsequent instalments of goods. Even express refusal to pay (as in this case, on incorrect legal advice) was held not to evince an intention to repudiate the contract.

'The law of England is that . . . payment for previous instalments is not normally a condition precedent to the liability to deliver — although it can be made so by express provision,' said Mr Justice Cooke in the New Zealand case of *Canterbury Pipe Lines Ltd* v. *Christchurch Drainage Board* (1979).

To apply this to the JCT 80 situation: failure or refusal to pay sums certified as due as interim payments will not in itself give the contractor the right at common law to cease to perform his obligations under the contract. His only remedies are to sue on the interim certificate and ask for summary judgment under Order 14 of the Rules of the Supreme Court; or to utilise the specific provision made in JCT 80 clause 28.1.1.

Can the contractor reduce his activities on site, go slow, or partially suspend work until he is paid?

It would appear that he can.

Suspension of the work by a contractor may in fact amount to a reaffirmation of the contract and not an intention to repudiate it.

In *F. Treliving & Co. Ltd* v. *Simplex Time Recorder Co. (UK) Ltd* (1981), it was argued before Judge Stabb QC that stopping work in the middle of a contract and refusing to go on was plainly repudiation and that for this purpose there was no difference between suspending and stopping.

The sub-contractor in that case had threatened to suspend work failing payment to him of monies he claimed were due for disruption, and the defendants had brought in another sub-contractor.

Judge Stabb said:

> 'I should have thought that "suspend" eliminates the essential quality for repudiation by refusal to go on and introduces a temporary quality into the stop.
>
> I have been referred to the cases of *Sweet and Maxwell Ltd* v. *Universal News Services Ltd* (1964) and *Woodar Investment Development Ltd* v. *Wimpey Construction (UK) Co. Ltd* (1980). Both of these are authorities for the proposition that a party who takes action relying upon the terms of a contract and not manifesting otherwise an intention to abandon the contract is not to be treated as repudiating, even if he is wrong in his construction of the contract.
>
> That is not really this case.
>
> Looking at the conduct of Treliving objectively, I ask myself the question whether it has been shown that it was their clear intention to abandon and to refuse performance of the contract.
>
> I think that the answer to that question is "no".
>
> They rightly contended that Simplex were in breach of the condition to afford them unimpeded access. Breach of a condition so fundamental to this sub-contract would have entitled Treliving to treat the sub-contract as repudiated.
>
> They did not do so, but gave Simplex the opportunity to which Simplex assented when [the Managing Director of Simplex] indicated his willingness to pay a substantial sum in part satisfaction of their claim.
>
> The failure of Simplex to do so caused, and in my view justifiably caused, Treliving to hold up performance of the work temporarily in order to give Simplex further time to meet that claim, the entitlement to which they had already acknowledged.
>
> At that time it is clear that it was the intention of Treliving to continue in due course the performance of the sub-contract.
>
> Treliving have not been shown to have repudiated the

> sub-contract, which was accordingly wrongfully terminated by Simplex.'

Similarly, in *Hill* v. *London Borough of Camden* (1980) contractors working under a JCT 63 contract reduced the size of the labour force on site and started removing from the site a number of pieces of equipment such as dumpers and concrete mixers apparently because they were dissatisfied because of delay in the architect's certification of extra work executed under variation orders.

The defendant council, the employers, claimed that those acts evinced an intention to repudiate this contract. Lord Justice Lawton said:

> 'It is impossible to say that they did anything of the kind. The one thing they do not purport to do was to leave the site and indeed the employers have never suggested that they did.
>
> Indeed their subsequent conduct indicates . . . that they were treating the contract as still subsisting. All that can be said against them is that by removing men and plant from the site in the way that they did they may not have been "regularly and diligently proceeding with the work".
>
> I see no reason why the defendants should say the plaintiffs have unlawfully repudiated the contract by what they did . . . The most they can say is that they may be able to prove some small amount of damage as the result of the activities of the plaintiffs at the material date. . . .'

With that Lord Justice Ormrod agreed:

> 'Everything the plaintiffs have done in this case has evinced not an intention not to be bound by the contract but the precise contrary and they have evinced an intention to treat the contract as still subsisting.'

Their intention to treat the contract as still subsisting was evinced, apparently, even by their giving notice to determine under what is now JCT 80 clause 28.1 [9.06].

Contractor's failure to proceed with due diligence

As will be seen, a default by a contractor in failing 'to proceed regularly and diligently with the works' is made the ground for premature determination of his employment: clause 27.1.2.

There is also, tucked away obscurely in clause 23.1, as if almost an after-thought, a contractual obligation that the contractor, once given possession of the site, 'shall thereupon begin the Works, regularly and diligently proceed with the same'.

It can be said at once that this obligation is a subsidiary term of the contract, a warranty not a condition. Breach of it will not entitle the employer at common law to treat it as repudiatory conduct, and if he chooses to sue for breach of the term, his damages would be merely nominal, for the simple reason that he has suffered no loss. The contractor's primary obligation is to complete the works by the contract date: how he does so is no concern of the employer and the employer's only entitlement to damages are those set out in the contract, namely liquidated damages for failure to complete on time.

It has been suggested by one writer that if there were no such express term in JCT 80, one would be implied: see *Hudson*, 10th Edition, page 611 *et seq.* and Supplement. The present writer does not agree. The contract is perfectly clear about the contractor's primary obligation: it is to do the work specified by the date for completion. Clause 23.2 is secondary and subsidiary surplusage. If it were not there, there would be no need to import it to give commercial effectiveness to the contract. The learned editor has apparently misunderstood both the doctrine of implied terms laid down in *The Moorcock* (1889) and also the true nature of lump sum building contracts.

Incidentally, nowhere in the JCT 80 contract does the contractor undertake that he will not 'wholly suspend the carrying out of the works before completion': reasonably, unreasonably, or at all.

Yet, by a piece of rather inept wording, this is described as a 'default' by JCT 80 clause 27.1. Whatever 'default' may mean, this is clearly not a breach of contract, since there is no specific requirement in all its eighty thousand words that the contractor will not, if it suits his convenience or method of working, wholly suspend the carrying out of the work.

An interesting situation arose when a contractor, working under the ICE Conditions on a section of the M25 motorway, secured an extension of time from the engineers of six months because of variations consisting of new concrete paving specifications and piling problems with a retaining wall, which brought the date for completion to 26 August 1985. With only half a kilometre to do of a 8.6 km stretch, the contractor wholly suspended all work in August 1984 and dismissed his labour force, intending to resume on 21 April 1985. As the managing director of the contractors said: 'Why do it when we can do it considerably cheaper after 21 April and be reasonably assured of warmer weather?' He was a prophet, for the severe weather of early 1985 caused other contractors on the same motorway to shut down for several weeks because temperatures were too low.

Although complaints were made in the House of Commons, the contractor was clearly within his rights: *GLC* v. *Cleveland Bridge and Engineering Co. Ltd.* (1984).

The work being executed on the Thames Barrier, to which that case referred, was also under the ICE Conditions. By clause 19, the employers were entitled to discharge the contractors if they failed to exercise 'due diligence or expedition' in the performance of the contract.

Mr Justice Staughton held that the totality of the contractor's obligation had been performed within the time limit of the contract and therefore the contractors were not liable for damages. He held that the provision for determination in clause 19 did not impose a contractual obligation to proceed with 'due diligence'.

Even if it had been a contractual obligation, his lordship said, 'due diligence and expedition' were to be interpreted in the light of the other obligations as to time in the contract.

> 'In this case the overall deadline has been complied with; and as it is a general principle of building contracts that it is for the contractor to plan and perform his work as desired within the contractual period, it cannot be said that they have failed to exercise due diligence and expedition.'

At issue were increased costs due to late manufacture of certain components and it was claimed that the contractors were disqualified from receiving the additional price because they had failed to proceed with due diligence.

With respect, his lordship was plainly right.

Similarly there is no contractual obligation under the JCT 80 contract that the contractor will not wholly suspend the contract works if it suits his convenience to do so. The obligation to proceed with 'due diligence', as has been seen, is a subordinate and subsidiary obligation, for breach of which only nominal damages could be recovered since the contractor's primary obligation is to complete the work in the contractual period. The employer has lost nothing if the work is suspended or the contractor fails to proceed with diligence, provided he gets the work completed by the due date.

A multitude of defects and the evident impossibility of completing by the contractual date may evince an intention to repudiate the contract.

In *Sutcliffe* v. *Chippendale and Edmondson* (1971), the case which eventually ended up in the House of Lords as *Sutcliffe* v. *Thackrah* (1973), it was argued that JCT 63 clause 25 set out certain matters for which an employer alone was entitled to determine a contract and that the employer's right to determine was limited to the matters set out in that clause and now substantially reproduced in JCT 80 clause 27.

Alternatively, it was argued that the words 'without prejudice to any other rights or remedies which the employer may possess' should be construed as reserving the right to the employer to determine the contract in the event of repudiation of it by the contractor but that such repudiation could not be anything less than the express examples of default contained in JCT 63 clause 25.

Judge Stabb in the Official Referee's court did not accept either argument. He held that the express reservation of 'any other rights or remedies which the employer may possess' preserved an employer's normal rights at common law and that if the contractor's conduct could properly be interpreted as amounting to a repudiation of the contract, or to a breach of a fundamental condition, the employer was entitled to treat the contract as at an end. He said:

> 'The whole combination of circumstances . . . did justify the plaintiff in ordering the contractors off the site.
>
> Their manifest inability to comply with the completion date requirements, the nature and number of complaints from sub-contractors, and [the] . . . admission that in May and June the quality of work was deteriorating and the number of defects multiplying . . . all point to the plaintiff's expressed view that the contractors had neither the ability, competence or the will . . . to complete the work in the manner required by the contract. Accordingly, I find that the plaintiff was justified in determining the contract.'

It will be observed that this issue arose in an action not between the employer and the contractor but between the employer and his own architects.

The original completion date for a house under a JCT 63 contract was 31 January 1964; without extensions of time and without deducting liquidated damages, the architects extended this to 27 March 1964. They then advised their client to accept a further variation of the terms of the contract by extending the completion date to 13 May 1964 with liquidated damages increased from £25 per week to £75. The work was not completed on the extended date and with effect from 27 June 1964 the employment of the contractor was determined by the plaintiff employers. The house was eventually completed by other contractors by approximately Christmas 1964.

It is also noteworthy that reliance was placed upon dicta of Lord Justice Holroyd Pearce in *Yeoman Credit Ltd* v. *Apps* (1962) to the effect that:

> 'an accumulation of the defects which taken singly might well have been within the exemption clause, but taken *en masse* constitute such a non-performance or repudiation or breach going to the root of the contract as disentitles the owners to take refuge behind an exception clause intended to give protection to those breaches which are not inconsistent with and not destructive of the whole essence of the contract.'

That case was decided before the doctrine of fundamental breach in relation to exemption clauses was disapproved by the House of Lords in the 1980 *Securicor* case but it may well be that under current law a multitude of minor breaches of contract may be treated as evincing an intention to repudiate the contract, even while purporting to perform it, or be such a breach as will discharge the other party from any further performance.

Gross malperformance of a contract will be deemed to be a repudiatory breach, as where a banner advertising a brand of peas was flown behind an aircraft during the two minutes silence once observed on Armistice Day: *Aerial Advertising* v. *Bachelors Peas* (1938).

9.02 Determination without breach

The JCT 80 contract in clauses 27 and 28 provides for determination of the contractor's employment (but not the contract) on the happening of events which are not repudiatory breaches of contract or, indeed, breaches.

The two clauses do not make them breaches of contract and in spite of the words 'without prejudice to any other rights or remedies which the Employer may possess' (clause 27) or 'the contractor may possess' (clause 28), neither in fact has any remedies other than those specifically conferred by the contract.

In *Thomas Feather & Co (Bradford) Ltd* v. *Keighley Corporation* (1953) Thomas Feather had contracted to build houses for the corporation under the 1939 RIBA Contract which provided, in terms similar to JCT 80 clauses 19.2 and 27.1.4, that the contractor's employment could be determined if, without consent, they employed sub-contractors to do any part of the work. Thomas Feather did sub-contract without consent, and the employers determined the contract.

The council then employed other contractors to finish the work at an extra cost of £21,000. An arbitrator awarded the council this sum as damages. The award was set aside.

Lord Goddard said:

> 'The meaning of this clause is clear enough. Here is a particular term of the contract which, ordinarily, would not be regarded as one going to the root of the contract, so that breach of it would not have given the corporation power to determine the contract or to treat that particular breach as a repudiator by the builder.'

But he held that, because the contract specifically provided that, in the event of unauthorised sub-contracting, the employment should be determined, 'it would not otherwise have been . . . but for that provision'. He held therefore that since there was no breach of contract, there could be no damages. The council's only remedy was the determination of the employment.

His lordship was surely in error in holding that the common law does not require a contractor to perform his obligations himself [4.14], unless it is immaterial who does 'a rough description' of work. He was also in error in holding that if the parties make any term 'of the essence of the contract', as they did, then it does not become a fundamental term of the contract, breach of which discharges the other party from any further performance and entitles him to damages.

But on his two false premises that employment of a sub-contractor was not a breach of the contract, he was undoubtedly right to hold that the only remedy conferred was determination of the contractor's employment.

9.03 Form of notices

Where a construction contract contains express provision for determination on default by a contractor, it used to be said that it was sufficient if notice of this was given in general terms, and without reference to the specific clause in the contract relied upon by the employer. But this now appears doubtful and a meticulous adherence to the exact terms of the contract is necessary.

Even if such a notice is invalid, it may be possible for an employer to rely upon the common law doctrine of repudiation. In *Supamarl Ltd* v. *Federated Homes Ltd* (1981) a dispute between Federated Homes Ltd (the main contractor for a large GLC housing development at Wakeham's Green, Crawley) and the ground work sub-contractor came before Judge Lewis Hawser QC sitting on Official Referee's business on preliminary legal points.

The ground work sub-contract contained provisions: 'All works to be completed to the satisfaction of the local authority and Federated Homes Ltd' and, 'Our detailed programme is to form part of this contract'.

The contract itself contained the following conditions:

> 'The contractor shall carry out and complete the works in accordance with this contract in every respect in accordance with the directions and to the reasonable satisfaction of the engineers.'

The same clause authorised the main contractors to employ other persons if the sub-contractor had not complied within seven days with any notice served by the engineer.

Clause 18 provided, 'If the contractor shall make default in any of the following respects, that is to say . . . If he fails to proceed with the works with reasonable diligence' and for determination of the contract in these events.

There were three other contracts providing for unit external works. The wording of these about determination of the contract was in different terms to the ground work contract. In these, clause 14 read:

> 'If in the opinion of the employer the sub-contractor shall fail to observe the conditions of this contract or shall fail to maintain a proper rate or standard of work the employer has the right to terminate the contract on demand, written or otherwise.'

'It is common ground,' said the judge, 'that clause 14 should be interpreted as if it included the words "based on reasonable grounds" after the words "if in the opinion of the employer". In other words, the right to terminate is not to depend upon any capricious or unreasonable view formed by the employer.'

The action arose out of a letter sent on 7 October 1976 which terminated the four contracts under the clauses quoted above. Supamarl claimed that Federated were not entitled to end their contracts, and counterclaimed that Federated had repudiated them. The judge said: 'It is common ground that if Federated's termination was not lawful, they were the repudiating party.'

The letter purporting to determine the contract was sent by recorded delivery instead of registered post as prescribed by the contract. But no point was taken on this and it was accepted that the notice was validly received.

In the course of the case it was suggested that Supamarl had slowed down because they were discontented with delays in payment for extra work. As to this, the judge said:

> 'Even if Federated were in breach of contract in failure to pay, I do not think that this would excuse or justify failure by Supamarl to proceed with their work in a proper manner.
>
> Looking at the picture as a whole and taking a broad view, I am satisfied . . . [Supamarl] had not for some time been proceeding with their works with due diligence or at a proper rate.'

It was submitted that the notice of determination did not comply with the requirements of the contract. On this the judge said:

> '[Counsel] for Supamarl accepted that such a notice may in appropriate circumstances be of a general character and that it need not specifically refer to clause 18.
>
> The former concession is grounded on the case of *Pauling* v. *The Mayor, Aldermen, and Burgesses of the Borough of Dover* (1855) and the case of *Hounslow London Borough* v. *Twickenham Garden Developments Ltd* (1970). In the *Pauling* case, the notice was in this form:
>
> "I give notice to you to supply all proper and sufficient materials and labour for the due prosecution of the works, and with due expedition to proceed therewith; and further, that if you shall for the space of seven days after the giving of this notice fail or

> neglect to comply therewith, I shall as engineer, and on behalf of the said A.B. [the defendants] take the works wholly out of your hands."

Baron Alderson, in the course of the argument, said this: "How could a notice be drawn to state more specifically that the contractor must proceed faster with the work on hand?"

Chief Baron Pollock said this: "We are all of opinion that the notice is sufficient. It would be a difficult matter to frame a notice specifically, unless the party giving it were bound to descend into minute particulars; which I think he cannot be called upon to do."

In the *Hounslow* case, Mr Justice Megarry (as he then was) took the same view. The notice in that case was in general terms, though it was in more formal language and it did specifically refer to the relevant contract clause.

[Counsel] for Supamarl argued that the notice here was bad because it was ambiguous. It did not clearly relate to the roads and sewers contract and it did not make it clear that if the default continued the power to forfeit or determine would be exercised.

Furthermore, he drew attention to one particular passage in the judgment in the *Hounslow* case. There the learned judge says this:

> "All that I think the notice need to do is to direct the contractor to what is said to be amiss: and this was plainly done by this notice. If the contractor had sought particulars of the alleged default and had been refused them, other considerations might have arisen. But as it stands, I hold that here too, there is nothing in the point."

Counsel submitted on the basis of this observation not that the notice of this was bad because of the absence of particulars, but that the effect of a request for particulars which he said occurred in the course of the correspondence in this case was that the contractor was not in default until they were furnished. In other words, he said that such a request could and did in this case operate to suspend the effect and operation of the notice.

I find myself unable to accept these arguments. I think the approach to a matter of this sort is well stated by Mr Justice John Stephenson (as he then was), in the case of *Goodwin & Sons* v. *Fawcett* (1965). At the very end of his judgment he said this:

> "However, the whole contract must be construed in a commonsense business way as a building contract, and this notice had been properly served on the true interpretation of the words used."

> In my judgment, the notice was *ab initio* valid and as this contractor was at that time in breach, the question of continuance depends simply on whether he remedied his default and not in the letter he wrote.'

He held that all four contracts had been properly determined and continued:

> 'In these circumstances, strictly speaking it is not necessary for me to deal with Federated's alternative submission that on any view Supamarl repudiated the four contracts, entitling Federated to accept repudiation and determine them which they did.
>
> I have already found that Supamarl were guilty of substantial breaches sufficient to bring into operation the termination provisions of the contract. Of course, it does not at all follow that such breaches would have constituted a repudiation which at common law would entitle Federated to treat the contracts as at an end.
>
> It is sometimes put as a refusal to perform something which goes to the root or the essence of the contract, or words or conduct which evince an intention to repudiate the party's obligations under the contract and no longer be bound by it.
>
> A party seeking to rely on repudiation implied from conduct must show that the party in default has so conducted himself as to lead a reasonable man to believe that he will not perform at the stipulated time or in the stipulated manner. . . . [He held that] If the contractual termination was ineffective, Federated were entitled to and did effectively terminate the contracts as having been repudiated by Supamarl.'

In spite of this judgment, it is submitted that if either the employer or the contractor wishes to determine the contract under JCT 80 clause 27 or 28 they should do so with the strictest regard to all the formalities required. In particular, it will be noted that for the purpose of clause 27 it is the architect and not his employer who has to give notice, but it is the employer who has to determine the contractor's employment, not the architect.

It is submitted that this judgment is unreliable in three particular respects: firstly, that the conduct relied upon as evidencing an intention to repudiate fell far short of that required by law; secondly, the machinery for premature determination was not correctly activated; thirdly, the judgment shows a lamentable confusion between determination of a *contract* by reason of repudiatory breach and determination of *employment* under the terms of the contract.

9.04 Not be given unreasonably

Both the employer's right to give notice determining the contract before completion and the contractor's right are subject to the words 'provided that such notice shall not be given unreasonably or vexatiously'.

The meaning of these words was discussed by Lord Justice Ormrod in *Hill* v. *London Borough of Camden* (1980). It was argued before the Court of Appeal that the contractor who had given notice required by what is now JCT 80 clause 28.1 had acted unreasonably when an architect's certificate for £84,518 was not paid and when the council expressly refused to pay it, claiming that they had an equitable set-off against the contractor on the principle of *Gilbert-Ash* v. *Modern Engineering* (1974), which entitled them not to pay. On a subsequent application by the plaintiffs for summary judgment under Order 14, the Master 'did not think much of that defence' and gave judgment for the full amount in the certificate.

> 'I find it difficult, in the circumstances of this case, [said Lord Justice Ormrod] to imagine how the plaintiff could be said to have behaved unreasonably in this regard. It seems to me that they had the most cogent reasons for taking advantage of such remedies as the contract gave them.
>
> But what the word "unreasonably" means in this context, one does not know.
>
> I imagine that it is meant to protect an employer who is a day out of time in payment, or whose cheque is in the post, or perhaps because the bank has closed or there has been a delay in clearing the cheque or something — something purely accidental or purely incidental so that the court could see that the contractor was taking advantage of the other side in circumstances in which, from a business point of view, it would be totally unfair and almost smacking of sharp practice.
>
> I can think of no other sensible construction of the word "unreasonably" in this context.'

Lord Justice Lawton for his part said:

'I am far from clear as to what conduct on the part of a contractor would make what he did unreasonable or vexatious. In my judgment, it was not unreasonable, for two reasons.

First, at the date when all this was happening any damage which the employers had sustained would have been minimal and secondly, they had no business to withhold money for work which had already been done and materials which had already been supplied.

The very essence of the provisions of the contract about payment on the architect's certificate was to maintain the cash flow of the contractors and when the cash flow is cut off without good reason it does not lie in the mouth of the employers to say that the plaintiffs were acting unreasonably.'

9.05 Determination for non-payment: JCT 80 clause 28.1

Normally time is not of the essence of payment in any contract and failure to pay one instalment or certified sum due does not amount to evidence of intention to repudiate the contract. The appropriate remedy for the unpaid contractor is to sue on the certificate while continuing the work.

However, JCT 80 clause 28.1.1 changes all that by stating that the contractor may determine if:

> 'the Employer does not pay the amount properly due to the Contractor on any certificate . . . within 14 days from the issue of that certificate, and continues such default for 7 days after receipt by registered post or recorded delivery of a notice from the Contractor stating that notice of determination under clause 28 will be served if payment is not made within 7 days from receipt thereof.'

No doubt the insertion of the word 'properly' before 'due' will result in the argument being raised that, in addition to the deductions expressly authorised by the contract, an employer is entitled to withhold unliquidated and unqualified sums as a counterclaim for alleged breaches of contract and to use them as an equitable set-off. The argument is entirely fallacious since by implication all claims to set-off, except such sums as are expressly made deductible, are by this contract excluded: see [6.16].

It is submitted that the only meaning of the words 'properly due' are the certified sums, less any deductions authorised by the contract.

'By registered post or recorded delivery': it is unclear whether these words are mandatory, in which case a delivery by hand to the employer will be excluded, or merely directory. In *Hill* v. *London Borough of Camden* (*supra*) the Court of Appeal, having considered the point, declined to decide it. The prudent contractor will therefore ensure that his notice is sent in the fashion indicated even if he delivers another by hand.

However, so far as the period named is concerned, the court did decide that the important provision in this clause was the word 'served'.

Lord Justice Lawton said:

> 'Service means that the notice has got to be given to the other party. It cannot be "served" on the other party until the other party has received it. It is after the other party has received the notice, the determination of the employment comes about.'

Lord Justice Ormrod said:

> 'On general principles, a notice to determine the employment of the contractor must operate from the moment when it is brought to the attention of the other party. . . . Such a notice operates only from the time it is received.'

9.06 Determination by contractor: JCT 80 clause 28

Much criticism has been directed at what is now JCT 80 clause 28 in that it allows the contractor to determine the contract and withdraw prematurely for matters which are substantially less than the repudiation by the employer that common law would require. 'A large number of the matters referred to in paragraph (c) are neither the legal nor the moral responsibility of the employer or his architect . . .' observes I. N. Duncan Wallace in his *Building and Civil Engineering Standard Forms* (1969). 'It will be seen that many of the events which giving rise to the right to determine are not even breaches of contract. Other events while they may be breaches, may not necessarily be repudiatory in effect,' comments Donald Keating in his *Building Contracts* (2nd Edition, 1978).

This clause is indeed extraordinarily onerous on employers and the consequences, even though they have to some degree been mitigated by JCT 80, are serious; so much so that although local authority and public employers have some sanctions (such as removal from the tender list) to restrain irresponsible determination by the contractor if he finds the contract is likely to be unprofitable, the private employer may have none and would be well advised to delete the whole clause and allow the contractor to rely on his common law rights.

Before considering the terms of this clause, it is proposed to note the alterations from JCT 63 clause 26, which itself was amended in July 1973.

JCT 63 clause 26(2) had a proviso that, in addition to all other remedies, the contractor on determination 'may take possession of and shall have a lien upon all unfixed goods and materials which may have become the property of the Employer under clause 14' (i.e. because the employer has paid for them in interim certificates). That has now been dropped, not out of any consideration for the employer but because JCT took the view that such a lien was 'without value' and 'might have been invalid unless registered as a charge', presumably under sections 95 and 96 of the Companies Act 1948.

The word 'properly' has now been inserted in JCT 80 clause 28.1, formerly JCT 63 26(1)(a). It is thought not to have any significance: a certified sum is due or it is not due.

The word 'Clause 22 perils' has been substituted for the risks formerly dealt with in JCT 63 clauses 20[A] or 20[B].

What has been retained has been the positive requirement for suspension introduced by the July 1973 amendment to JCT 63, which requires the period to be stated and does not provide that, in the absence of such indication, the periods shall be one month and three months. If no periods are specified there are no periods after which JCT 80 clause 28 can operate and the provision is ineffective.

9.07 Determination for interference with certificates: JCT 80 clause 28.1

Lord Somervell in *Burden* v. *Swansea Corporation* expressed the hope that the words used in the earlier RIBA contract would be clarified. His wish has not been granted: they remain as they have done for many years. The contractor may determine if:

> 28.1.2 'the Employer interferes with or obstructs the issue of any certificate due under this Contract;'

In the *Burden* case it was held that these words applied if the employer directed the architect (including his employee architect) to withhold a certificate or dictated the amount in it. Any interference with an architect's duty to act independently and fairly between the parties will be sufficient grounds and if the architect disqualifies himself by such conduct in accepting such instructions, the contractor can sue without a certificate: *Hickman* v. *Roberts* (1913); *Panamena Europa* v. *Leyland* (1947). An instruction to an in-house architect to withhold a certificate until local authority auditors have approved it is clearly caught by this clause.

9.08 Suspension of the works: JCT 80 clause 28

The contractor is entitled to give notice to determine the contract if the work is suspended for the periods named in the Appendix. By a footnote, of no contractual significance, it is suggested that the period should be three months in respect of JCT 80 clause 28.1.3.2 relating to clause 22 perils, and one month in respect of all the rest.

'Civil commotion' occurs twice. It is a clause 22 peril and it is also a separate ground under JCT 80 clause 28.1.33. If the advice in the footnote is followed the contractor can determine his employment if there is 'loss or damage' to the works by civil commotion after *three months*' suspension, or if the work is suspended for *one month* by reason of 'civil commotion' without loss or damage to the works. Is this really what the draftsmen intended?

> 28.1.3 'the carrying out of the whole or substantially the whole of the uncompleted Works (other than the execution of work required under clause 17) is suspended for a continuous period of the length named in the Appendix by reason of:
>
> .3.1 force majeure; or
>
> .3.2 where either clause 22A or clause 22B applies loss or damage to the Works (unless caused by the negligence of the contractor, his servants or agents or of any sub-contractor, his servants or agents) occasioned by any one or more of the Clause 22 Perils; or
>
> .3.3 civil commotion; or
>
> .3.4 Architect's instructions issued under clause 2.3, 13.2 or 23.2 unless caused by reason of some negligence or default of the Contractor; or

.3.5 the Contractor not having received in due time necessary instructions, drawings, details or levels from the Architect for which he specifically applied in writing provided that such application was made on a date which having regard to the Completion Date was neither unreasonably distant from nor unreasonably close to the date on which it was necessary for him to receive the same; or

.3.6 delay in the execution of work not forming part of this Contract by the Employer himself or by persons employed or otherwise engaged by the Employer as referred to in clause 29 . . . or

.3.7 the opening up for inspection of any work covered up or the testing of any of the work, materials or goods in accordance with clause 8.3 (including making good in consequence of such opening up or testing), unless the inspection or test showed that the work, materials or goods were not in accordance with this Contract;

then the Contractor may thereupon by notice by registered post or recorded delivery to the Employer or Architect forthwith determine the employment of the Contractor under this Contract; provided that such notice shall not be given unreasonably or vexatiously.'

It is only suspension of the work under the architect's powers under JCT 80 clause 2.3 (discrepancies), 13.2 (variations) or clause 23.2 (postponement of the works) which entitles the contractor to determine under JCT 80 clause 28.1.3.4.

But clause 28.1.3.6 may apply to the not infrequent delays caused by statutory undertakers, if *Boot* v. *Central Lancashire New Town Development Corporation* is right [1.09].

In all, three months and one month seem far too brief periods for the considerable consequences that may follow.

9.09 Consequences of determination by the contractor: JCT 80 clause 28.2

The contract provides for what is to happen if the contractor exercises his right to determine the contract. He will remove himself and his plant from the site: JCT 80 clause 28.2.1. This appears to deal with the problem raised by *Hounslow LBC* v. *Twickenham Garden*

Developments (1971) where the contractor refused to go and it was held that the employer was not entitled to an interim injunction to remove him [9.01].

But without prejudice to other accrued rights or remedies the contractor is then entitled to the following:

28.2.1 'the total value of work completed at the date of determination . . .

.2.2 the total value of work begun and executed but not completed at the date of determination, the value being ascertained in accordance with clause 13.5 as if such works were a Variation required by the Architect under clause 13.2 . . .

.2.3 any sum ascertained in respect of direct loss and/or expense under clause 26 . . .

.2.4 the cost of materials or goods properly ordered for the works . . .

.2.5 the reasonable cost of removal under clause 28.2.1;

.2.6 any direct loss and/or damage caused to the Contractor or any Nominated Sub-Contractor by the determination.'

The deadly result of the last provision is that the employer has to pay the contractor and every nominated sub-contractor the whole of the profits they would have made if the contract had been carried to completion.

That is the implication of the case of *Wraight Ltd* v. *P. H. & T. (Holdings) Ltd* (1968).

Work started under a JCT 63 contract in November 1964 and because of difficult soil conditions was suspended shortly afterwards on the instructions of the architect, presumably under clause 21(2). The contractors withdrew and never returned. On 1 February 1965 they served notice of determination on the employers under clause 26(1)(c)(iv), the period of suspension (unspecified in the case) having elapsed.

The matter came before the High Court on a case stated in an award by an arbitrator. The full details are not set out in the judgment but Mr Justice Megaw (as he then was) said that the issue was the meaning of what is now JCT 80 clause 28.2.2.6.

The contract provided for a 10 per cent 'establishment charge and profit' on the builder's work and 12½ per cent 'for overheads', totalling in all £5159. 'That sum . . . was thus something which [the contractors] were entitled to receive as part of their total remuneration from [the employers] if the contract was duly performed,' said the judge.

The contractors claimed before the arbitrator the damage flowing directly from the determination of the contract under JCT 63 clause 26(2)(b)(vi) (now JCT 80 clause 28.2.2.6), that is damages flowing naturally in the usual course of events which, they alleged, included the loss of gross profit, i.e. the £5159.

The employers claimed that this was not 'direct loss and/or damage' which they contended meant direct, certain, out-of-pocket loss or damage in contrast to consequential loss. The £5159 was not such direct loss because it was (a) uncertain in that it was dependent on uncertain events, i.e. whether the profit would have been made, and (b) excluded possible alternative uses of the contractor's facilities.

Moreover, they contended that determination under JCT 63 clause 26 should not be treated as a breach of contract by the employer since it was determination under the contract and without any breach or other fault by the employer. Why then should it be equated to damages for breach of contract?

Mr Justice Megaw said:

> 'There are no grounds for giving the words "direct loss and/or damage" any other meaning than that which they have . . . in a case of breach of contract. The [contractors] are, as a matter of law, entitled to recover that which they would have obtained if this contract had been fulfilled . . .
>
> [They] are entitled to recover, as being "direct loss and/or damage", those sums of money they would have made if the contract had been performed.'

9.10 The employer's rights to determine: JCT 80 clause 27

By comparison with the contractor's rights, which are precise and advantageous to the contractor, those of the employer are nebulous and elusive.

The reasons for determination because of the conduct of the contractor are as follows:

27.1.1 'if without reasonable cause he wholly suspends the carrying out of the Works before completion thereof; or

.1.2 if he fails to proceed regularly and diligently with the Works; or

.1.3 if he refuses or persistently neglects to comply with a written notice from the Architect requiring him to remove defective work or improper materials or goods and by such refusal or neglect the Works are materially affected; or

.1.4 if he fails to comply with the provisions of either clause 19 or 19A.'

The operative word in the first ground is 'wholly'. The employer has no right unless there is a complete cessation of work on the part not only of the contractor himself but of any sub-contractor. Moreover, 'reasonable cause' can be any matter of any kind and is not limited to those contemplated in JCT 80 clause 25.

A number of cases have been concerned with failure to proceed 'regularly and diligently' with the works and all indicate the great difficulty in satisfying the court about this. The arbitration regarding Peterborough Hospital, which appears in the *Guinness Book of Records* as the longest lasting (239 days), was concerned largely with this single point.

Failure to comply with an intended programme, especially one that is not a contractual document, is no evidence, since the contractual obligation is to complete within a specific period, and there is great scope for the contractor to show that even if he has failed to proceed 'regularly and diligently' there are reasons for which he is not to blame.

So far as JCT 80 clause 27.1.3 is concerned, written notice must mean one which complies fully and in detail with clause 8.4. In *Holland Hannen & Cubitt* v. *Welsh Health Technical Services Organisation* (1981), all the windows installed by nominated sub-contractors in the Rhyl hospital let in water. Judge John Newey QC, sitting on Official Referee's business, dealt with the notices 'or purported notices' given by the architects to the contractors under JCT 63 clause 6(4) which was in the same terms as the current provision. The notices were dated, he said, 8 May 1974, 7 February 1975 and 30 January 1976.

'The first of these documents was in letter form and did not mention clause 6(4). The second and third were more formal and referred to the sub-clause. None of them, however, instructed Crittalls to remove the window assemblies from the site and the

> effect of so doing would have been to expose the interior of the hospital to wind and rain.
>
> In the absence of any oral testimony from the architects I have no hesitation that their motives for issuing the notices were declaratory only . . .
>
> An architect's power under clause 6(4) is simply to instruct the removal of work or materials from the site on the ground that they are not in accordance with the contract.
>
> A notice which does not require the removal of anything at all is not a valid notice under clause 6(4). The three purported notices in this case were all invalid and of no effect under the contract.'

Further, he held that even if he was wrong about that, the architects had waived the employer's rights under the contract in that the instructions had not been complied with. The sanction, he held, under the contract was to bring in other persons, after notice, to remove them under JCT 63 clause 2(1), now JCT 80 clause 4.1.2.

'Since [the architects] did not invoke their sanction under clause 2(1) of bringing in other contractors to remove the windows, then effect has long since been spent,' the judge said.

But there are other sanctions for non-compliance with JCT 80 clause 8.4, including withholding payment under clause 30.2.4, and that provided in clause 27.1.3 for determination, subject to the proviso that 'by such refusal or neglect the works are materially affected'. If that reasoning in the preceding paragraph is correct, presumably this sanction will also effectively be waived if not exercised within a reasonable period. But does the architect have power under clause 8.4 to order the contractor to remove the work of a nominated sub-contractor? NSC/4 clause 29.1.3 deals with sub-contractors' defective work.

The reference in JCT 80 clause 27.1.4 to clause 19 is, of course, to the prohibition against assignment of the contract and reference has already been made to the difficulty of placing meaning to clause 19.1.

Clause 19A is the one that appears only in local authority editions and deals with the payment of rates of wages, hours and conditions of labour established for the trade or industry in the district where the work is being carried out. This clause is based on the Fair Wages Resolution of the House of Commons in 1946, passed at a time when it was considered that one obligation of good government was to ensure that local and public authorities employed only contractors who would pay proper wages to their staff. A local authority who sought to exercise the powers conferred by this clause to determine a contract would receive little sympathy from the courts, which have already indicated their view in *Racal Communications Ltd* v. *The Pay Board* (1974) that this clause must be regarded as obsolete. The present Government have since repudiated this obligation and the country's obligation under treaty to the International Labour Organisation.

If a contractor's employment is determined under this section, the architect is under an obligation to ascertain and certify 'any direct loss and/or damage' caused to the employer: JCT 80 clause 27.4.4.

9.11 Procedure for employer's determination: JCT 80 clause 27.1.4

In the unlikely event of the employer being able to establish to the satisfaction of a court any of the events set out above, the procedure must then be strictly adhered to: that is, the architect must give the contractor by registered post or recorded delivery notice specifying the default alleged.

It was pointed out by the Court of Appeal in *Hill* v. *London Borough of Camden* (1980) that the employer himself has no power to send this notice and that it will not be effectual if he does; also that the provision regarding registered post or recorded delivery 'might be construed as mandatory by the court', so that a notice handed to the contractor or delivered by other means would not be a valid one.

Thereafter 'if the Contractor either shall continue such default for 14 days after receipt of such notice . . .' the employer may take action.

The operative period of this, in spite of the provision about the manner in which notice is to be served, runs only from the time when the notice is actually received by the contractor. Needless to say, proof of continuance of default is absolutely essential and difficult to obtain in a form which cannot be disputed.

After the fourteen day period, the employer this time has ten days, and ten days only 'after such continuance or repetition . . . by notice by registered post or recorded delivery forthwith [to] determine the employment of the contractor'. Clearly notice from the architect will not suffice.

From this it is obviously extremely difficult, if not impossible, for any employer to successfully determine a contractor's employment on any of the grounds set out in clause 27.1. A notice to determine which fails will be construed as an unlawful repudiation of the contract, with all the inevitable consequences.

9.12 The contractor's insolvency: JCT 80 clause 27.2

The contract provides for the automatic determination of employment of the contractor if a receiver is appointed or the contractor becomes insolvent or passes a resolution for winding-up. Notice is not necessary. Normally the position of a receiver is that he is bound by the contractual obligations of the company to which he is appointed: *Parsons* v. *Sovereign Bank of Canada* (1912). As Lord Justice Edmund Davies said in *George Barker* v. *Eynon* (1974) about a receiver:

> 'He must fulfil company trading contracts entered into before his appointment or he renders it liable to damages if he declines to do so . . . neither the receiver nor the debenture holders are in any way relieved by the former's appointment from the obligation which, by pre-appointment contracts, the company had undertaken.'

It would be better, therefore, for an employer if the contract did not provide for automatic determination in the event of the appointment of a receiver by the debenture holders.

Receivers regularly 'hive down' all the assets of a company to which they are appointed to a new company, leaving the one of which they are receivers and managers with no assets. They can therefore repudiate contracts with impunity since the original company has no assets.

However, it is probable that they could be restrained by injunction from repudiating wrongfully a construction contract, and by a *Mareva* injunction from disposing of the assets pending an action for breach of contract. They might well also be liable for actions which could be brought against them personally, i.e. for the torts of procuring a breach of contract and conspiracy. It is therefore extremely foolish to provide for automatic determination.

By contrast a liquidator normally has the option of adopting or disclaiming any onerous contract: Companies Act 1948, section 323(1). The provision in the contract for automatic determination appears to remove this option from him, since there is no subsisting employment for him to determine.

Where there has been a winding-up order or a resolution for winding-up, the employer loses one of the rights conferred by the contract on the determination, namely the right to have assigned to him all agreements relating to the supply of goods: JCT 80 clause 27.4.2. This right still prevails if a receiver is appointed and it is therefore a wise course for the employer or the architect (either can exercise this power) to call for such assignments.

9.13 Determination for corruption: JCT 80 clause 27

JCT 80 clause 27.3 in the local authorities edition, but not in the private edition, provides:

> 'The Employer shall be entitled to determine the employment of the Contractor under this or any other contract, if the Contractor shall have offered or given or agreed to give to any person any gift or consideration of any kind as an inducement or reward for doing or forbearing to do or for having done or forborne to do any action in relation to the obtaining or execution of this or any other contract with the Employer, or for showing or forbearing to show favour or disfavour to any person in relation to this or any other contract with the Employer, or if the like acts shall have been done by any person employed by the Contractor or acting on his behalf (whether with or without the knowledge of the Contractor), or if in relation to this or any other contract with the Employer the Contractor or any person employed by him or acting on his behalf shall have committed any offence under the Prevention of Corruption Acts 1889 to 1916 or shall have given any fee or reward the receipt of which is an offence under sub-section (2) of section 117 of the Local Government Act 1972 or any re-enactment thereof.'

The clause is apparently wide enough to cover the operation of public relations consultants, so-called, or other work-procuring agents acting on a contractor's behalf.

This clause is only an empowering one: it does not require the employer to determine the contract in these circumstances.

There is no provision for corruption the other way round, i.e. corruption of the contractor by a servant or agent of the employer.

9.14 Employer's rights on determination: JCT 80 clause 27.4

On determination either by the employer by notice, or automatically under JCT 80 clause 27.2, the employer is entitled to employ other persons to carry on the works and may for that purpose use 'all temporary building, plant and tools etc.' adjacent to the works: JCT 80 clause 27.4.1.

If so required, the contractor is required to assign to the employer within fourteen days of determination the benefit of any agreement for supply of goods etc. to the works, subject to conditions: JCT 80 clause 27.4.2.1.

The employer can pay any unpaid supplier or sub-contractor directly — except where the contractor is already in liquidation — and may deduct such sums from any money due to the contractor: JCT 80 clause 27.4.2.2.

The contractor, if required, must remove plant (presumably only where a liquidator has not been appointed) and if he does not do so within a reasonable time the employer can 'sell any such property of the contractor': JCT 80 clause 27.4.23. The question as to whether the employer can give any purchaser a title to goods so sold does not seem to have been considered by the draftsmen of the contract.

Finally, JCT 80 clause 27.4.4 provides that:

> 'the Contractor shall allow or pay the employer in the manner hereinafter appearing the amount of any direct loss and/or damage caused to the Employer by the determination. . . .'

The manner prescribed exonerates the employer from making any further payments to the contractor, and requires the architect to certify 'within a reasonable time' the amount of expense incurred by the employer and 'any direct loss and/or damage caused to the Employer' and such certified sums constitute debts due from the contractor to the employer.

In the event of the insolvency of the contractor, the bankruptcy rules regarding set-off contained in the Bankruptcy Act 1914, section 31 apply to limited companies: *Mersey Steel and Iron* v. *Naylor Benson* (1882). The employer is therefore entitled to set off monies due to the contractor against his own expenses and 'direct loss and/or damage' certified by the architect under this clause.

This leaves the situation regarding monies due to nominated subcontractors or suppliers without contractual provision. In the event of the contractor's insolvency the employer now loses the right to pay them direct. Since such payments are to be treated as money due to be paid to the contractor, as a debt from the employer, presumably all such sums are capable of being set off against the employer's claim for expenses and 'direct loss and/or damage'.

In view of *Wraight Ltd* v. *P. H. & T. (Holdings) Ltd* (1968) it may be taken that this clause entitles the employer to the whole of the loss he has suffered, as damages due to the determination, including his loss of liquidated damages, making good defects, employing new contractors at a higher price and such consequential losses which are reasonably foreseeable as arising out of what is to be treated as equivalent to the contractor's breach of contract. There can rarely be any sum due to be paid to a liquidator.

Moreover, the right of set-off in insolvency is much more extensive than the right of set-off given by statute by the Supreme Court of Judicature Act 1925, section 39(1), or equity. There is a right to set off amounts owing on other contracts, or any other business dealing between the parties whether such sums are liquidated or not, whether they arise in contract, tort or otherwise and whether they are immediately due or not: *re D. H. Curtis (Builders) Ltd* (1978). It follows that the employer is entitled to set off his claim against an insolvent contractor before the architect has certified his damages under JCT 80 clause 27.4.4.

Chapter 10

Claims and counterclaims: disputes and their resolution

10.01 Contractors' claims against the employer

As has been pointed out, JCT 80 purports to be a lump sum contract. In spite of that there are commonly additional sums payable to the contractor, for example, by way of fluctuations or variations. In fact, the only JCT contract which has ever been known to come out at the contract sum was that for the renovation of All Souls' Church in Langham Place, London and that may justly be regarded as a miracle of divine grace.

Contractors' claims against the employer for money are commonly divided by the industry into two sorts: contractual and extra-contractual (or 'ex-contractual', as the industry terms them).

In fact there are two other categories as well: quasi-contractual and *ex gratia*.

Contractual claims are those which arise out of the express provisions of the contract, and they can be sub-divided into:

(a) claims for extra work, including that comprised in variation orders
(b) claims arising from the provisions of the contract for the contractor to be indemnified by the employer for the 'direct loss and/or expense' which he suffers as the result of certain events.

The rather quaintly-named 'extra-contractual' means that the claims are outside the terms of the contract. In reality they are no more than claims for damages for breach of contract.

As such they may be allegations that the express, or implied, terms of the contract were broken by the employer or his agent, the architect, for whom he is responsible within the area of that agent's express and implied authority.

If such claims are pursued before a court, a contractor can seek relief other than the payment of damages. He may seek a declaration, an injunction which can be mandatory or prohibitive, or an order of specific performance, all of which have potential as the foundations of monetary claims.

The case of *Holland Hannen & Cubitt* v. *Welsh Health Technical Services Organisation and Ors* (1981) is a good illustration of the possibilities of such actions. The main contractor under a JCT 63 contract for the erection of a hospital at Rhyl sued the employers, the nominated steelwork sub-contractors, Redpath Dorman Long Ltd (trading as CED Building Services) and Crittall Windows Ltd. The first defendants brought in as third parties their structural engineers and their architects. In separate actions the structural engineers sued Shockcrete Products Ltd (in liquidation) and Costain Concrete Co. Ltd, designers and fabricators of precast floor units, and Redpath Dorman Long Ltd also sued Shockcrete Products Ltd and Costain Concrete Co. Ltd regarding the structural steelworks. The three cases, which all arose out of alleged defects in the hospital were in due course consolidated and then sub-divided into a number of subtrials which related to specific defects. The first sub-trial, concerning the windows, engaged four Queen's Counsel and their juniors (and counsel for other parties were present as observers) from November 1980 until February 1981 before judgment running to sixty-three pages in the transcript was given on 28 April 1981. After that most of the parties settled the cases, but there was a subsequent hearing about defective flooring. In all, the case involved Judge John Newey sitting for 156 days and it was not finally concluded until 1984.

The facts are mentioned since they are highly relevant to the arbitration clauses in the JCT 80 and other construction contracts [10.11].

In this place, however, it is relevant to look at the claims which Cubitt's made against their employers, who started off as the Welsh Hospital Board but who were on 1 April 1974 replaced by the named defendants, the Welsh Health Technical Services Organisation.

The contractor's action against the employer which started in 1977 with a High Court writ alleged four categories of claims.

Firstly, there were claims for declarations:

(a) that architect's instructions issued under JCT 63 clause 6(4) (now JCT 80 clause 8.4) for removal of work were void as not being in accordance with the contract;

(b) that Cubitt's were entitled on or about 21 April 1976 to a variation order (for remedial works);

(c) that they were entitled to an order on or about 21 April 1976 postponing the works.

Secondly, there were claims for monies due under the contract and for loss and expense incurred and payable by the employers. That is, for what has been described above as 'contractual claims'.

Thirdly, the cause of action was for damages for breach of contract i.e. an 'ex-contractual' claim in the jargon of the industry.

Fourthly, 'damages for negligent misrepresentation and/or breach of warranty and/or pursuant to the Misrepresentation Act 1967 arising out of representations made or warranties given by or on behalf of [the Welsh Health Technical Services Organisation]'.

It is not proposed to elaborate on this aspect of the matter since these allegations were dependent on alleged pre- and post-contractual observations about the nature of the work, items included in the bills of quantities under the heading 'sequence of operations' and about the design made by the employer and/or the professional agents to the contractors, some before the JCT 63 contract was signed, others later.

It should, however, be noted in passing that such items as 'sequence of operations' in a bill of quantities are capable of providing evidence of misrepresentation not only where they misdescribe the work but also where they inadequately describe the necessary sequence or do not describe it at all while purporting to do so.

What is, however, of interest is what the plaintiffs, Cubitt's, alleged against the employers in their statement of claim. It was claimed under *The Moorcock* doctrine [2.06] that there were implied terms in the contract that the employers, their architects, agents and nominated sub-contractors would do all things necessary to enable the contractors to carry out and complete the works 'expeditiously, economically and in accordance with the main contract'.

There is express authority that normally when an architect nominates a sub-contractor he does not do so as agent of the employer so as to make the employer liable to the contractor for the delays or defects of the sub-contractor: *Leslie* v. *Metropolitan Asylums* (1901); *Mitchell* v. *Guildford Union* (1903). In this case, however, the Department of Health and Social Security selected the nominated sub-contractors, notified the employers who were to be used as window design-and-install contractors, agreed with the sub-contractors after the contracts had been signed a variation in design, and the employers in turn instructed their architects to make the nomination.

The other implied term was that neither the employer, the architect nor the sub-contractor would 'in any way hinder or prevent' the contractor from carrying out and completing the works.

Breaches of the express terms of the JCT 63 contract alleged were as follows:

(a) Although practical completion took place on 31 March 1978, 'in breach of Condition 15(1) of the main contract the architects failed to certify practical completion for that date'. In consequence, Cubitt's suffered loss and damage by having to maintain a site establishment between 1 April and 20 December 1978.
(b) The architects failed, in breach of what was termed 'Condition 27(d)(ii)', to certify the dates on which the nominated sub-contractors ought reasonably to have completed.
(c) The architects had issued a clause 6(4) notice requiring the removal of all floor levels which were outside the tolerances laid down on CP 204 Part II (Metric) and all affected work by following trades. The contractors challenged this under JCT 63 clause 2(2) but carried out remedial works. They claimed the cost on the ground 'that there was no defect such as would entitle the architect to issue an instruction under JCT 63 clause 6(4)', now JCT 80 clause 8.4; and even if they were defective, this was because of the design. They claimed that they were entitled to instructions about the floor under JCT 63 clause 11(1) or 21(2) and that the architect refused to give such instruction.

It should however be observed that if an architect issues notice under JCT 80 clause 8.4 he has no obligation to issue instruction to the contractor as to what he has to do, because clause JCT 80 clause 2.1 (JCT 63 clause 1) requires the contractor 'to carry out and complete the works in compliance with the contract documents, using materials and workmanship of the quality and standards therein specified'. In normal circumstances the suggestion that, if an architect issues a notice condemning any part of the work he then has to issue instructions to the contractor as to how he should fulfil the obligations of his contract, cannot be supported; still less the plaintiff's contention that any work that had to be carried out to make the works comply with the contract had to be a variation order.

The normal principle of law is that, in the absence of any express term of the contract to the contrary, a contractor is responsible to the employer for the work of a sub-contractor, even a nominated one, for all the sub-contractor is doing is providing vicarious performance for the contractor. But here it was being alleged by the contractor that it was not the workmanship that was at fault but the design, which had been prepared by the nominated sub-contractor and approved by the architect.

The contractor also pleaded that on the facts stated he was entitled to an extension of time under JCT 63 clause 23(e) 'by reason of architects' instructions issued under clauses 1(2), 11(1) or 21(2) of these Conditions' (now JCT 80 clause 25.4.5), which the architect had not issued, and under JCT 63 clause 23(f) (now JCT 80 clause 25.4.6) by reason of the contractor not having received in due time necessary instructions. It is also pleaded that loss and expense caused thereby is payable under JCT 63 clause 24(1)(a) (now JCT 80 clause 26.2.1) — failure to receive instructions; and 24(1)(e) (now JCT 80 clause 26.2.5) — architect's instructions to postpone works: instructions which he had not issued.

The pleadings concluded with the allegations that the contractors were entitled to extensions of time and loss and expense, as indicated above, and also as an alternative that 'time was at large' and they were not liable in consequence to liquidated damages for failure to complete on time.

These pleadings are summarised as an excellent example of what 'extra-contractual' claims can be raised against an employer.

The trial judge, Judge John Newey QC, in the windows sub-trial held that Cubitt's were not entitled to an extension of time 'by reason of delay on the part' of the nominated window sub-contractors, Crittall's.

He also held that the architects, on finding the windows defective, 'because of JCT 63 clause 3(4) and an implied term in the main contract' should have issued a variation order. And that if the window sub-contractor did not indemnify the employer, the architects should be liable to them for failure to issue a variation order and 'in failure to grant an extension of time'.

In claims made (however quantified) for what are in essence unliquidated damages for breach of contract — these so-called 'ex-contractual' claims — the architect has neither express nor implied authority from the employer to agree to them, either as to liability or amount. It may be that on occasions he is expressly authorised by his client to negotiate with the contractor such claims, but such authority does not normally arise out of the terms of the JCT contract, his own conditions of engagement or the relationship of architect and employer.

The architect certainly has no authority from the employer to deal with claims by the contractor for quasi-contractual payments, that is for a *quantum meruit* [2.16], still less for *ex gratia* payments.

Contractual and ex-contractual claims, however, have one thing in common — the claim must flow directly and in the natural course of things from the event which gives rise to it, and not merely as something without which the loss would not have been incurred. There can be no case of 'for want of a nail, the shoe was lost; for want of a shoe the horse was lost; for want of a horse the rider was lost; for want of the rider — the battle was lost'. The event must be the direct and substantial cause of the loss and not merely the occasion of it.

In any well drawn contract a change of expression would indicate a change of meaning and, therefore, the words 'loss and/or damage' in JCT 63 clause 26(vi) and JCT 80 clause 28.2.2.6 should in the ordinary way mean something different from 'loss and/or expense' in JCT 63 clause 24(1) and JCT 80 clause 26.1.

Moreover, where the word 'expense' is used as well as 'loss', in the ordinary way of construction of a contract, it should mean something other than loss.

However, it appears that 'loss and/or expense' must be equated to the common law right to damages and in particular to damages for breach of contract.

As Mr Justice Megaw (as he then was) observed in *Wraight* v. *PHT* (1968):

> 'There are no grounds for giving to the words "direct loss and/or damage caused to the Contractor by the determination" any other meaning than that which they have, for example, in a case of breach of contract or other question of the relationship of a fault to damage in a legal context.'

The same applies to the words 'loss and/or expense'. These must be assessed in the same way as damages at common law for breach of contract. In particular, the claimant must have taken reasonable steps to mitigate his loss:

> 'In assessing the damages for breach . . . a jury will, of course, take into account whatever a prudent man ought to have done whereby his loss . . . would have been diminished': *Frost* v. *Knight* (1872).

Further, damages are also subject to the 'foreseeable test' set out in *Hadley* v. *Baxendale* (1854), *Victoria Laundry* v. *Newman* (1949), and *East Ham Borough Council* v. *Bernard Sunley* (1966); and the contractor is only entitled to recover such part of the loss actually resulting as was *at the time of the contract* reasonably foreseeable as liable to result from the breach.

This principle alone normally precludes a contractor from recovering loss and/or expense on other contracts he may have in hand.

Finally, the loss and/or expense must have been *caused* by the breach and not merely be the occasion for it: *Weld-Blundell* v. *Stephens* (1920). It must be a *causa causans* and not a *causa sine qua non*, some precedent event but for which a *causa causans* would not have operated. If a burglar is chased across roof tops by a policeman and the policeman falls through a skylight and breaks his neck, the burglar's acts are a *causa sine qua non.* If the burglar pushes the policeman off the roof his act is a *causa causans.*

A building firm, in breach of contractual and common law obligations, failed to provide a workman with suitable equipment to carry out his trade. He therefore made use of improvised equipment and was injured as a result. The Court of Appeal held that his injuries were not caused by his employer's failure to provide equipment. That was a *causa sine qua non*, not a *causa causans*: *Quinn* v. *Burch Bros* (1966). The cause of his injuries were the defects of his own improvised equipment.

10.02 Valid claims?

Very frequently contractors' claims are based on a fallacious view of causation. In one case, an architect refused to include in an interim certificate work which he considered defective. The contractor put in a claim for loss and expense, including overdraft interest, alleging that there had been delay and disruption by the architect in failing to provide funds to allow the contractor to proceed. The failure to receive cash may well have delayed the contractor; but, quite apart from any other consideration, the architect's interim certificate was no more than a *causa sine qua non*. It was the contractor's lack of adequate funds that was the *causa causans*, if the delay and disruption was proved. It mattered not whether the architect's refusal to certify was justified or non-justified — it did not cause the loss and expense.

Just such an example is to be found in the widely read and respected *Building and Civil Engineering Claims* (2nd Edition) by R. D. Wood.

He suggests that if the architect issues the instruction:

'*Omit* Red facings PC £100 per thousand to be obtained from Messrs A.B.C.
Add Yellow facings PC £70 per thousand to be obtained from Messrs X.Y.Z.'

this entitles the contractor to make a very considerable claim. The following situation is presented.

The variation order is issued a month before the facing brickwork is to start. This is followed by a strike by A.B.C.'s workmen which prevents the facing bricks from being supplied until the strike is over. Under pressure from the architect, and with the help of the builder, a promise of early deliveries is obtained from another firm X.Y.Z. However, deliveries prove unreliable, starting three weeks after they are needed and arriving in 2000 brick loads a day instead of 8000. The bricks themselves prove to be badly burnt, of irregular sizes beyond the tolerances permitted under the British Standard for bricks, and thanks to careless loading an unacceptable proportion are damaged.

Mr Wood then states that the variation was the cause of the loss and that the surveyor should notify the architect of the position and request further instructions. He then goes on to detail the costs and claims which, in his view, would arise:

(a) The builder had allowed for only one unloading of bricks from the supplier's lorry. The further unloadings that were required must therefore be paid for by the supplier.
(b) The supplier must also bear the builder's cost of loading defective bricks back on to the supplier's lorry.

(c) The architect, having ordered the faulty bricks to be sorted, must pay for this and will then contra-charge the supplier in the monthly certificate (depending however on the terms of the nominated contract and/or agreed exclusions).
(d) If the bricks were delivered bonded and had to be opened up before loading, the extra cost would be paid by the architect in the same way as described in (c).
(e) Delay caused by the late delivery of the bricks will merit an extension of contract but no payment by the employer to the contractor, since there is no contract between the employer and nominated supplier. This cost will, if possible, be claimed from the manufacturer's supplier, depending on what the contract allows.
(f) The irregular size of the bricks will increase the cost of laying the bricks and may require additional mortar. This will give grounds for a star item and rate under Clause 11(4)(b) of JCT 63.
(g) If it can be established that the architect was late in nominating the supplier, a claim for damages from the employer could perhaps be made under the express or implied terms of clauses 11(3) and 24(1) of the JCT form.

With respect, I must take issue with this argument. The loss and expense which the contractor suffered in no way flow from the architect's variation order, except as a *causa sine qua non*. It was not caused by the variation order. What caused the contractor's loss? Clearly, it was the breach of contract by X.Y.Z.

The sale of the yellow facings by X.Y.Z. to the contractor was a sale subject to the Sale of Goods Act 1893 (now the Sale of Goods Act 1979) and there were three breaches of sections 13 and 14 in that the goods delivered were not of merchantable quality, did not meet the description and were not reasonably fit for the purpose required.

To suggest that this variation order was the cause of the loss cannot be accepted. The sole cause of the loss was the brick manufacturer's breach of contract.

Moreover, the contractor had the option of rejecting the whole delivery, since it was clearly a breach of conditions of the contract between the contractor and the brick manufacturer. He failed to mitigate his loss.

If the contractor chose to accept the goods and treat the breaches of conditions as breaches of warranties, he cannot possibly have a claim against the employer, since he is the author of his own misfortune. He will, however, be fully entitled to recover all that as against the supplier.

Moreover, although it is possible that some naive architect might negligently fall into the baited trap laid for him by the contractor asking him what to do with defective bricks delivered, an experienced architect will know that he is not there to tell the contractor how to do his work, that his authority as agent does not extend to that, and will tell the contractor to get on with his job and not waste his time with frivolous applications for instructions.

The prudent architect will not insist on inspecting future deliveries of bricks. He will advise the contractor that the obligation rests on him to do work that is reasonably fit for the purpose required and if brickwork proves to be defective, whatever the cause, a notice condemning it will be issued under JCT 80 clause 8.4 (formerly clause 6 of JCT 63).

All these principles were clearly set out as long ago as 1956 in *Kirk and Kirk* v. *Croydon Corporation*.

10.03 The relationship of JCT 80 clause 25 to clause 26

One of the fallacies prevalent in the industry is that an extension of time under JCT 80 clause 25 results in 'prolongation of the contract' and these facts entitle the contractor to money as a result.

To emphasise that this is not so, the commentary in this book on JCT 80 clause 25 and JCT 80 clause 26 are widely separated. JCT 80 clause 25 entitles the contractor to relief from paying liquidated damages at the date named in the contract. It does not in any way entitle him to one penny of monetary compensation for the fact that the architect has extended the contractor's time for completion. He is not entitled to claim even items set out in 'Preliminaries' for the extended period.

Once again, I must take issue with Mr Wood. In his *Supplement to Building and Civil Engineering Claims* at page 176 (and elsewhere) he refers to 'Disruption for matters under clause 25.4.1 to .12 unless paid for under another provision'. He then goes on to show, in his worked 'Example of a Contractual Claim for Prolongation of a Contract', that because the architect has extended time for completion by 48 weeks, of which 37 are to be the owner's responsibility, on the basis of Preliminaries set out in SMM6 clause B.4.1, the employer must pay a proportionate rate for the cost of one theodolite, four levels, twelve ranging rods and two chainmen involved in setting out.

But JCT 80 clause 25 entitles the contractor to nothing except an extension of time, which relieves him from liquidated damages.

Although JCT 80 clause 26 follows clause 25 and makes use of some, but by no means all, of the relevant events, there is no connection between them. An architect can extend time under clause 25 without giving any 'loss and/or expense' under clause 26 and he can give 'loss and/or expenses' under clause 26 without granting an extension of time under clause 25. There is no connection, logically or contractually, between them.

10.04 What does 'direct loss and/or expense' mean?: JCT 80 clause 26

Regrettably, it cannot be assumed in a JCT contract that the same words bear the same meaning in different clauses throughout the contract and, conversely, a change of words does not always convey a change of meaning.

In JCT 80 clause 26, however, the meaning is restricted even more because the contractor is not entitled to all the damages which may have flowed from the events set out in the clause. It is limited by three things: the regular progress of the work must have been affected by the event, it must have been affected 'materially', that is to say to a substantial extent, and it must be something for which there is no other payment. The pricing of variations, for example, under JCT 63 includes elements for overheads and profits.

The earlier RIBA contracts provided for the contractor to recover loss and/or expense beyond that provided for in, or reasonably contemplated by, the contract. JCT 63 changed that completely, but the industry continues to read the earlier words as an accurate expression of these later ones. They are not. The contractor is not entitled to compensation under JCT 80 clause 26 because any of the matters there set out cause him extra work, or greater expense in doing his work or more difficulty than he contemplated. He will not be entitled to compensation for loss of productivity.

It is only if the 'regular progress of the works . . . has been materially affected' that the contractor can be compensated.

If the regular progress of the works is not materially affected, e.g. because the contractor has time in hand or has sufficient slack in his programme, the happening of a relevant event entitles him to no extras. Even if the contractor has been put to extra expense, if the 'regular progress of the works has not been materially affected' he is entitled to nothing extra. Those words are a condition precedent to payment and architects are advised to require explicit evidence as to exactly how 'the regular progress of the works has been materially affected' before authorising any payment under this clause.

This, needless to say, is not a popular view with contractors — see Vincent Powell-Smith and John Sims: *Building Contract Claims* pages 124–152.

It is a misconception to believe that clause 26 in any way provides 'financial reimbursement for specified breaches of contract by . . . the employer himself or . . . others for whom he is responsible'. The architect is in no way empowered under JCT 80 to decide whether the employer has been in breach of contract, whether the contractor has suffered damages or still less to quantify them; and it is certainly no part of his duty as an independent professional.

Under clause 26, in certain circumstances only, the contractor is entitled to extra payment, i.e. 'direct loss and/or expense . . . for which he would not be reimbursed by a payment under any other provision in this contract'.

The addition of the word 'direct' seems, in the view of most commentators, to add little or nothing to the words 'loss and/or expense'. However, there is a definition of the word in *Saintline* v. *Richardson, Westgarth & Co.* (1940) by Mr Justice Atkinson:

> 'Direct damage is that which flows directly from the breach without intervening cause and independently of special circumstances, while indirect damage does not so flow.'

The judge drew a distinction between those damages which were 'direct' and those that were 'consequential damages', meaning those which were not the direct and natural result of the breach of contract: see also *Millard Machinery* v. *Dawbay & Co.* (1935).

10.05 Interest as 'direct loss and expense'

Great difficulties arise as a result of the decision of the Court of Appeal in *Minter* v. *Welsh Health Technical Services Organisation* (1980).

The House of Lords long ago held that interest is not payable on debts or as damages for failure to pay money on time: *London Chatham and Dover Railway* v. *South Eastern Railway* (1893). That indeed has been a rule of the common law as long as there has been a common law, and it was recently followed: *Smith Hogg* v. *Black Sea & Baltic* (1979).

Since then the House of Lords has reaffirmed this fundamental principle of the common law in *President of India* v. *La Pintada* (1984). Damages cannot be given for failure to pay money on the due date; any interest which a court can award must be based on statute or must be provided for by contract.

As a result, it was necessary for a statute, the Judgment Act 1838, to be introduced to authorise courts to award interest from the date of judgment until the date of payment.

Section 3(1) of the Law Reform (Miscellaneous Provisions) Act 1934 was necessary to enable the courts to award interest between the date when 'the cause of action' accrued and the date of judgment. Moreover, it applied only to 'courts of record', with the result that under the Act the county courts, not being courts of record, had no power to award interest from the time when money was due. The position regarding county courts was altered by the Administration of Justice Act 1982.

The Court of Appeal in the *Minter* case changed all that and held that 'financing charges', or interest, were payable as an element in 'loss and/or expense' under JCT 63 clause 24.

But it held that 'financing charges' were only payable for the interval between the time when they were incurred and when a claim was made under JCT 63 for re-imbursement. This led some commentators to point out that daily application would therefore have to be made to secure such interest.

It is interesting to see what reasons the court gave for disregarding the provisions of the common law.

Lord Justice Stephenson said:

> 'I do not think that today we should allow medieval abhorrence of usury to make us shrink from implying a promise to pay interest in a contract, if by refusing to imply it, we thereby deprive a party of what the contract appears, on its natural interpretation, to give him. There should be no presumption in favour of an anomaly and an anachronism.'

Lord Justice Ackner said:

> 'What the appellants here are seeking to claim is not interest on a debt, but a debt which has, as one of its constituent parts, interest charges which have been incurred.'

These reasons seem to have been vitiated by the *President of India* case (1984).

It will be seen that the judges in the Court of Appeal allowed interest, or rather 'financing charges', as an element of 'loss and/or expense' really because they disliked the decision of the House of Lords in the *London Chatham and Dover Railway* case *supra*. It is by no means explicit in the words 'loss and/or expense' that it included interest which a common law assessment of damages would have disallowed.

The House of Lords' reaffirmation of the old principle of the common law in the *President of India* case in 1984 must throw doubt upon the judgment of Mr Justice Kilner Brown in *Rees and Kirby Ltd* v. *Swansea Corporation* (1983).

In that, the judge joined in condemnation of the *London Chatham and Dover Railway* case, and said:

> 'Although I may privately agree with Lord Justice Stephenson that the case is an anomaly and an anachronism, it is not open to a judge of first instance to disregard its application. The most he can do is to seek properly to find a way round it. . . .'

However, Mr Justice Kilner Brown was, he confessed, unable to find a way round it and held 'with much regret' that he could not decide that the defendants were liable to £206,629 interest as damages for non-payment. But in the glorious intoxication of intellectual rebellion against the House of Lords' judgment by which he was bound, he decided that the contractors were entitled to this sum as interest under clauses 11(6) and 24(1)(a), (c) and (d) of a JCT 63 contract which was for a fixed sum of £845,301. The contractors were exactly a year late in handing over the housing estate on 4 July 1974. Before that date there had been variations which were valued, certified and paid in the final certificate.

Claims for 'loss and/or expense' under clauses 11(6) and 24(1) were not made until 17 July and 10 August 1979, more than five years after practical completion on 4 July 1974, and were in due course certified by the architect and paid by the employer.

How such a claim could have been accepted within the terms of JCT 63 clause 24 is hard to understand. The Judge did not specify how five years could be regarded as 'written application . . . within a reasonable time of it becoming apparent that the progress of the works has been affected'.

The judgment of Mr Justice Parker at first instance in *Minter* is greatly to be preferred to that of the Court of Appeal, being logical and demonstrating a clear understanding of JCT 63.

As Lord Justice Slade said in the course of argument in the Court Appeal in *Rees and Kirby* v. *Swansea Corporation* (1985):

> 'Financing charges are not a direct loss or expense. . . . By their very nature, they arise not because of a variation or a relevant event under clause 24 but because of the delay inevitable in reimbursing the contractor. . . . It is not a direct loss . . . because it flows not from the variation or event but from inevitable delay in payment. It is therefore too remote.'

Lord Justice O'Connor observed:

> 'A contractor cannot recover financing charges for the interval between the time when he does work and the time when he is paid on an interim certificate. There is no difference between that situation and claims for extra reimbursement under clauses 11 and 24' [of JCT 63].

The clear implication is that financing charges, alias interest, are not recoverable as 'direct loss and/or expense' under JCT 63 clauses 11 and 24.

However, in the end, the Court of Appeal allowed the appeal in the *Rees and Kirby* case only as to *quantum*. As this book went to press, both sides have indicated that they intend to appeal to the House of Lords. Leading counsel for the Council conceded that the court, in spite of its expressed views indicated in argument, was bound by the *Minter* decision, and with this the court, not surprisingly, agreed. He did not take the point, apparently, that the courts were precluded from opening up and reviewing the architect's certificate as to loss and expense under the principles laid down in the case of *Northern Regional Health Authority* v. *Derek Crouch* (1984).

The decision in the *Minter* case also appears to violate the House of Lords' decision in *The Edison* (1933). If 'financing charges' were incurred, it could only have been because the claimants had insufficient funds of their own to finance the contract. This seems to violate Lord Wright's principle:

> 'The appellants' actual loss, in so far as it was due to their impecuniosity, arose from that impecuniosity as a separate and concurrent clause, extraneous to and distinct in character . . . the impecuniosity was not traceable to the respondents' acts and . . . was outside the legal purview of the consequences of these acts. The law cannot take account of everything that follows a wrongful act. It regards some subsequent matters as outside the scope of its selection because "it were infinite to trace the causes" or the consequences of consequences.'

But this decision of the House of Lords has long been ignored by the courts, if it has ever indeed been followed: *Dodd* v. *City of Canterbury* (1980).

10.06 Condition precedents: JCT 80 clause 26

The wording of JCT 80 clause 26 is different from JCT 63 clause 24. Now the contractor has to apply as soon as 'the regular progress of the Works or any part thereof . . . is likely to be materially affected'.

It will now be difficult for even the most sympathetic Court of Appeal to hold that contractors are entitled to interest from the time they are likely to suffer loss or incur expense.

This is a condition precedent. It requires two things: (a) a written application and (b) that this should be made in the time specified by the contract. If both these are not fulfilled the architect not merely has a duty to reject any application, but will be guilty of negligence if he does not do so. This excludes the usual claims for prolongation of the contract submitted before or about the time of the final certificate.

10.07 Entertaining applications under JCT 80 clause 26

The architect is entitled to reject out of hand any application by the contractor which does not comply with this condition precedent. If he decides to entertain the claim, he should ensure that the contractor complies with the requirement of JCT 80 clause 26.1.2 to provide the architect with 'such information as should reasonably enable the Architect to form an opinion', together with quantified details of 'such loss and/or expense as is alleged'.

It is now clearly contemplated that there must be separate written applications in respect of each and every incident, with the cause of the claim specifically identified as one of the matters set out in clause

26. The practice of lumping together all incidents in one 'prolongation claim' is quite wrong.

It should be noted in particular that there is no possibility of a claim where the architect has extended time for completion for JCT 80 clause:

25.4.1	*force majeure*
25.4.2	exceptionally adverse weather conditions
25.4.3	clause 22 perils
25.4.4	strike or lock-out
25.4.7	delay on part of nominated sub-contractors or nominated suppliers
25.4.8.2	delay caused by supply of employer's own goods
25.4.9	statutory powers
25.4.10	Inability to get labour or materials
25.4.11	statutory undertakers carrying out statutory obligations.

In fact, it is well to bear in mind the observations of Judge Fay QC in *Boot* v. *Central Lancashire New Town*:

> 'The broad scheme of these provisions is plain. There are cases where the loss should be shared, and there are cases where the loss should be wholly borne by the employer.
>
> There are also cases, which do not fall within either of these conditions and which are the fault of the contractor, where the loss of both parties is wholly borne by the contractor, but in the cases where the fault is not that of the contractor, the scheme clearly is that in certain cases the loss lies where it falls.
>
> But in other cases the employer has to compensate the contractor in respect of delay, and that category should clearly be composed of cases where there is fault upon the employer or faults for which the employer can be said to bear some responsibility.'

It is against this background that what are termed 'Matters' in JCT 80 clause 26 (by contrast with the expression 'Relevant Events' in clause 25) should be considered.

10.08 'Matters' in JCT 80 clause 26

The first matter mentioned is that of the contractor not receiving in due time necessary instructions: JCT 80 clause 26.2.1. It should be observed here that such failure may be a breach of contract in that it is an implied term (it is not so expressly stated) that the architect will issue all necessary instructions in reasonable time for the contractor to do the work.

But the architect has no power to allow damages for that unless two conditions have been fulfilled: namely there must have been (a) a written application and (b) within the period specified by the contract.

Opening for inspection: clause 26.2.2. Here it is not operative where the work has been prematurely covered up before inspection by the architect or building inspector.

Discrepancy between contract drawings and bills: clause 26.2.3. This, of course, is subject to the requirement that the contractor who discovers such a discrepancy must *immediately* advise the architect: JCT 80 clause 2.3. If the contractor does discover such things and fails to notify them he is in breach of contract, cannot recover under this heading and is liable in damages.

Execution of work by persons engaged by employer: clause 26.2.4. This may relate to statutory undertakers.

Postponement of work: clause 26.2.5. As has been observed earlier, this cannot cover failure to give possession of the site.

Failure to give ingress or egress: clause 26.2.6.

Architect's instructions re variation: clause 26.2.7. The power to extend time for completion is contained in clause 25.4.5. The principle is, of course, that if there is extra work the architect must have power to extend the contractor's time for completion, otherwise time for completion will be at large and the liquidated damages irrecoverable. And it follows in this case that the contractor must receive not only the cost of the variations but also compensation for interruption to the progress of the works, if such can be proved. Although not expressly stated, in determining this as an issue of fact, the architect is entitled to take into account variations that amount to omissions and which have saved the contractor time. He is also entitled to consider whether the contractor is in breach of any of the terms of the contract, e.g. regarding defective work which has taken time to rectify.

When the architect has received all necessary information, he should then consider:

(a) What is the matter alleged in respect of this specific claim?
(b) Is it one where the employer (or the architect) is in some way at fault or responsible?
(c) Has the regular progress of the work been affected by the matter alleged?

(d) Has the regular progress of the work been *materially* affected?
(e) If so, has the contractor suffered direct loss and expense or merely consequential?
(f) Is it loss or expense for which he would not be reimbursed by a payment under any other provision of the contract?

10.09 Claims for site conditions

The general law is that the employer does not warrant to a contractor that the site is fit for the works designed for it or that the contractor will be able to construct the building on the site to that design: *Appleby* v. *Myers* (1867). If therefore the contractor meets difficulties which he did not anticipate he is entitled to no extra payment for them: *Bottoms* v. *York Corporation* (1892).

The position is however altered so far as the JCT contracts are concerned by what is contained in SMM6. It will be required that the bills are to have been prepared in accordance with this: JCT 80 clause 2.2.2.1.

To start with, A1 of the General Rules of SMM6 provides:

> 'The Standard Method of Measurement provides a uniform basis for measuring building works and embodies the essentials of good practice but *more detailed information than is required by this document shall be given where necessary in order to define the precise nature and extent of the required work*' [author's italics].

This seems to impose an obligation on the employer to provide the contractor with any information in his possession regarding site conditions when they are known to be difficult.

SMM6 specifically in D3.1(a) requires the employer to provide information about the ground water level, the level when various excavations have been carried out and accurate details as to periodical changes if any.

By D3.1(c) the employer has to provide 'Particulars of underground services'. But the most demanding provision is that contained in D3.2:

> 'If the above information is not available a description of the ground and strata *which is to be assumed* shall be stated' [author's italics].

This information, if not fully accurate, will undoubtedly entitle the contractor under JCT 80 clause 2.2.2.2 to treat corrections as a variation order for 'any error in description or quantity'.

Moreover, if the bills do not accurately set out 'the quality and quantity of the work' the contractor will be entitled to rely on JCT 80 clause 14.1.

In addition, there may be ex-contractual claims for misrepresentation as in the Australian case of *Morrison-Knudson International* v. *The Commonwealth* (1972) where information provided by the employer failed to give reference to the presence of cobbles in the clay subsoil.

Where the contractor has any design responsibility for foundations there may be collateral warranty by tender documents that the soil is as described, as in *Bacal Construction* v. *Northampton Development Corporation* (1975) where the tender described the soil as a mixture of Northamptonshire sand and upper lias clay, and tufa was subsequently discovered.

Moreover, section 8 of the Unfair Contract Terms Act 1977 restricts the power of an employer to exclude or disclaim liability for misrepresentation. There may also be liability in negligence on the principle of *Hedley Byrne* v. *Heller* (1963) for inaccurate information provided about site conditions by an architect or an employer.

10.10 Employer's rights in respect of defects

The JCT 80 contract is very detailed on the contractor's right against the employer but less specific regarding the employer's rights against the contractor, particularly in relation to defective workmanship.

The defects liability period and its purpose and implication are considered later [11.04].

The contractual position regarding defects discovered as the work is in progress is less clear: see also [11.01].

When such work is seen by the architect as it is being carried out, the position is tolerably clear and in practice creates little difficulty. The architect has power to order the removal of the defective work: JCT 80 clause 8.4. If Judge Newey QC was right in *Holland Hannen & Cubitt* v. *Welsh Health Technical Services Organisation* (1981), it is not enough for an architect purporting to act under this clause to condemn the work as defective. One would think it would have been sufficient where he specifically referred to the exercise of his powers under that clause. But the learned judge held that reference to the clause was insufficient and the architect has to spell out in actual words 'remove from the site the work'. He said: 'a notice which does not require the removal of anything at all is not a valid notice,' under JCT 63 clause 6(4).

However, in normal matters, clearly it is unnecessary for the architect to issue further instructions to replace the work when it is a matter of workmanship or materials. The contract requires the contractor, by JCT 80 clauses 2.1 and 8.1, to comply with the contract.

Difficulties arise when the defects are solely a design fault or when they are, as is not infrequent, a fixture of design fault and workmanship. After much controversy, Judge Newey held that the 1500 windows that let in water at the Rhyl Hospital did so because of the sealant, which he held to be a design fault on the part of the nominated sub-contractors for which the architects, who had approved the aluminium windows without enquiring what sealant was to be used, were also liable. Another judge might well have held that defective sealants were exclusively a failure of workmanship and materials. In the circumstances, the judge held that the architects' employers were liable to the contractor for the architects' failure to issue a variation order — a conclusion which was possibly reached more easily since the same architects, faced with the same faults in windows in a hospital to the same design at Gurnos, had issued a variation order to different main contractors.

Under JCT 80, contractors can claim to have no liability in respect of the design element in nominated sub-contractors' design work. The terms are contained in JCT 80 clause 35.21:

> 'Whether or not a Nominated Sub-Contractor is responsible to the Employer in the terms set out in clause 2 of the Agreement NSC/2 or clause 1 of the Agreement NSC/2a the Contractor shall not be responsible to the Employer in respect of any nominated sub-contract work for anything to which such terms relate. . . .'

Those sub-contract terms relate to 'the design of the sub-contract works', 'the selection of materials and goods for the sub-contract works' and 'the satisfaction of any performance specification'.

To what extent this clause will serve to exonerate the contractor for work not reasonably fit for the purpose required is doubtful in view of the current attitude of the courts regarding implied terms imposed by operation of law [4.03]. The Model Conditions for mechanical and electrical work, in clause 22, have a much more emphatic exclusion of liability clause, but it did nothing to exonerate the main contractors from liability for design and fabrication by nominated sub-contractors in *IBA* v. *EMI and BICC* (1980).

In JCT 80, clause 35.21 is further qualified by the words:

> '. . . Nothing in clause 35.21 shall be construed so as to affect the obligations of the Contractor under this Contract in regard to the supply of workmanship, materials and goods.'

It will be noted that there is no power under the contract for the architect to suspend work pending an investigation, apart from his power under clause 23.2 which deals with the postponement of work yet to be executed. The provision for 'opening up' etc. contained in JCT 80 clause 8.3, with all the financial problems that follow, is also inadequate. The situation in *Gloucestershire County Council* v. *Richardson* (1969), where defects in concrete columns from a nominated supplier, which were installed by the contractor, enabled the contractor to escape from the contract, is unlikely to be repeated in view of the changed attitude of the courts, no doubt as a result of the influence of the Law Commissioners' recommendations contained in Law Com. 95 regarding *Implied Terms in Contracts for the Supply of Goods* and the Supply of Goods and Services Act 1982.

10.11 Arbitration: joinder of parties: JCT 80 Article 5

Disputes between the employer and the contractor have to be submitted to arbitration: JCT 80 Article 5. But if either party should sue in the courts, unless the other seeks and secures an order staying the proceedings in accordance with the Arbitration Act 1950, the action will proceed.

One of the defects of arbitration, as opposed to litigation, is that an arbitrator can only be appointed as the result of an agreement between the parties. In the case of the JCT 63 contract, the arbitration could only be between the contractor and the employer. There was no way, without their consent, that the architect, any sub-contractor or supplier could be joined as a party to the arbitration.

The contractual situation is as shown in Figure 5.

In litigation the plaintiff is at liberty to sue whom he likes as defendants, whether he is in contractual relationship with them or not. In

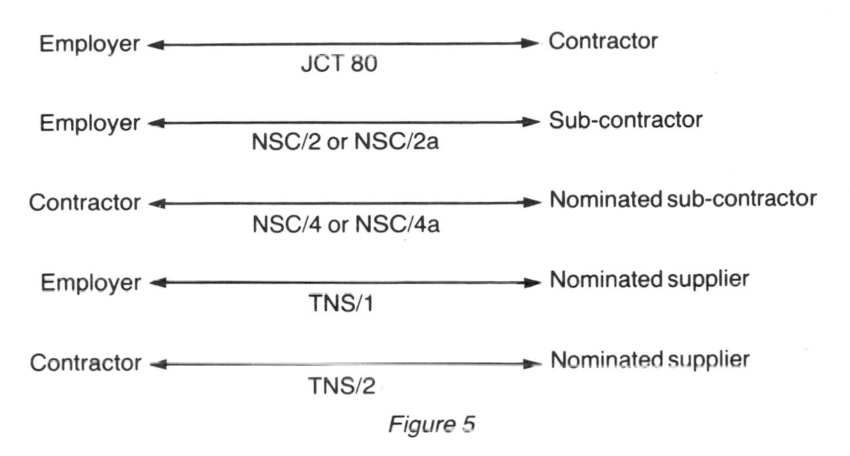

Figure 5

turn, a defendant can join anybody as a third party with the leave of the court on the basis that if the defendant is liable to the plaintiff, he is entitled to an indemnity or damages or a contribution under the Civil Liability (Contribution) Act 1978. A typical third party notice reads:

> 'TAKE NOTICE that this action has been brought by the Plaintiff against the Defendants. In it the Plaintiff alleges . . . etc, . . . The Defendant claims against you . . . to be indemnified against all damage, loss or expense resulting from the relief sought and against the costs of this action, and/or contribution under the Civil Liability (Contribution) Act 1978 on the grounds that . . .'

Here follows the usual statement of claim.

JCT 80 attempts to remedy the defect which normal arbitration procedure demonstrates in this respect by making contractual provision in all the different contracts that the parties will join in the arbitration, if the dispute or difference referred to arbitration under the contract raises issues which are substantially the same as issues raised in a related dispute.

In fact the new JCT 80 gives the employer a choice between the simple arbitration procedure between himself and the contractor, or the revised and extended procedure set out in JCT 80 Articles 5.1.4 and 5.1.5.

The option is exercised by the deletion of the words in the Appendix, 'Articles 5.1.4 and 5.1.5 apply'.

It is further provided that 'the Arbitrator shall have power to make all such directions and all necessary awards in the same way as if the procedure of the High Court as to joining one or more defendants or joining co-defendants or Third Parties was available to the parties and to him': JCT 80 Article 5.

The *JCT Guide* says that, as a result, 'the Arbitrator has the same powers as are available in the High Court for the joining of parties in legal proceedings'. This is over-optimistic. Nothing can compel any party to join an arbitration unless he is under a contractual obligation to do so. This is not the place to discuss the provision in detail and for this reference should be made to the present author's *Arbitration: Principles and Practice*.

The employer cannot, for example, bring in the architect or the structural engineers as third parties.

What is more, any of the parties who are not parties to the initial arbitration can object if they consider that 'the Arbitrator appointed to determine the dispute is not appropriately qualified'. This is a licence to refuse to join in since it is not subject to any test of reasonableness nor is any machinery provided for resolution of any dispute about the qualifications of the arbitrator.

It is an interesting attempt to make arbitrations more effective but it is doubtful whether it will be successful.

10.12 Arbitration: subject matter of dispute: JCT 80 Article 5

Another point of interest in the new arbitration clause contained in Article 5 is that it does not cover all disputes between the employer and the contractor.

Some are expressly excluded, e.g.: fair wages: JCT 80 clause 19A; 'the lump': JCT 80 clause 31.9; the VAT agreement, Supplemental Provisions clause 3.

Others are excluded because the Article now mentions only such matters as may be submitted:

Article 5

5.1 'In case any dispute or difference shall arise between the Employer or the Architect on his behalf and the Contractor, either during the progress or after the completion or abandonment of the Works, as to

5.1.1 the construction of this Contract, or

.2 any matter or thing of whatsoever nature arising hereunder or in connection herewith including any matter or thing left by this Contract to the discretion of the Architect or the withholding by the Architect of any certificate to which the Contractor may claim to be entitled or the adjustment of the Contract Sum under clause 30.6.2 or the rights and liabilities of the parties under clauses 27, 28, 32 or 33 or unreasonable withholding of consent or agreement by the Employer or the Architect on his behalf or by the Contractor, but

.3 excluding any dispute or difference under clause 19A, under clause 31 to the extent provided in clause 31.9 and under clause 3 of the VAT Agreement

then such dispute or difference shall be and is hereby referred to the arbitration and final decision of a person to be agreed between the parties to act as Arbitrator, or, failing agreement within 14 days after either party has given to the other a written request to concur in the appointment of an Arbitrator, a person to be appointed on the request of either party by the President or a Vice-President for the time being of the Royal Institute of British Architects.'

JCT 80 clause 5.1.1 is a historic relic as a reaction to an inaccurate *obiter* observation by a judge to the effect that an arbitrator had no power to construe the contract under which he was appointed, although the courts have held that it is a discretionary ground for not staying proceedings that a question regarding the construction of a contract arises by reason of the dispute: *Charles Osenton* v. *Johnston* (1941).

10.13 Arbitrations 'shall not be opened': JCT 80 Article 5.2

The meaning of these words is obscure. When is an arbitration 'opened'? There appears to be no authority as to what this means, although there is ample authority as to the actual date when there is a 'submission' to arbitration. Does it mean, as has been suggested, the time when the hearing, if any, starts?

The better view is that an arbitration 'opens' when there is a submission. It is to be regretted that, although the attention of the JCT has been called to the ambiguity of this word, they have not seen fit to remove it. It is submitted that what this is intended to say is that there shall be no submission to arbitration of a dispute, other than those specified, before practical completion or any of the other events specified in JCT 80 Article 5.2.

10.14 Arbitrations before completion: JCT 80 Article 5

The provision, which has been the subject of judicial criticism, that all arbitrations must wait until after 'Practical Completion' or 'alleged Practical Completion' or 'termination' or 'alleged termination of the Contractor's Employment' or 'abandonment of the works' is repeated: JCT 80 Article 5.2.

It is quite unnecessary for the Article to state that the contracting parties can agree to waive this provision. Of course they can. Any contract can be varied by agreement of the parties. It is noteworthy that the architect in this instance is authorised to vary the contract on the 'employer's behalf'. It is doubtful whether that lies within his power under his conditions of engagement or otherwise but no doubt an employer who has signed JCT 80 would be estopped from denying that his architect had this authority, by virtue of this Article.

The matters regarding which there can be an arbitration before practical completion, or other event set out in JCT 80 Article 5.2, are:

Article 3: Appointment of new architect or supervising officer
The contractor is given the right to object except where the new architect or supervising officer is an official of the local authority.

Article 4: Appointment of new quantity surveyor
The contractor is given the right to object except where the new quantity surveyor is an official of the local authority.

Clause 5.2
Early arbitration can also take place if there is a dispute (JCT 80 clause 5.2):

- 'whether or not the issue of an instruction is empowered by the Conditions', or
- 'whether a certificate has been improperly withheld', or
- 'whether a certificate is not in accordance with the Conditions'.

The second of these must be construed as meaning that the architect has entirely failed or refused to issue a certificate which he was under an obligation to do. It can have no application where the contractor does not like what is in the certificate.

The words 'whether a certificate is not in accordance with these Conditions' in JCT 63 clause 35(2) were said by I. N. Duncan

Wallace to open 'the door . . . to disputes on all matters which should be dealt with on interim certificates . . . and, as has been seen, these are very numerous': *Building and Civil Engineering Standard Forms* (1974), page 184.

That was not the view taken by the Court of Appeal in *Killby and Gayford Ltd* v. *Selincourt Ltd* (1973). Lord Denning there said:

> 'So long as a certificate is good on the face of it and is within the authority given by the contract, then it is "in accordance with the Conditions". It must be honoured.
>
> I do not think it is open to the employers of the contractors to challenge an interim certificate by saying it is too much or too little or includes this or omits that or that the extras were not sanctioned in writing. Such matters must be left till after the practical completion of the work.
>
> There are no grounds whatsoever for saying that this certificate is not in accordance with the contract. It must be honoured.'

This was a case of an employer wishing to dispute an interim certificate but the principle is equally applicable to a contractor who wishes to do so. The court was specifically concerned with JCT 63 clause 35(2) and was unanimous in its interpretation of it. It also rejected the argument based on *Gilbert-Ash* v. *Modern Engineering* (1973) that an equitable right of set-off entitled an employer not to pay a sum certified under the Standard Form of Building Contract. Lord Justice Megaw said: 'There is no evidence . . . that the certificate is not in accordance with the contract.'

With respect, the Court of Appeal was plainly right. There can be an early arbitration as to whether 'the certificate is in accordance with the conditions'. There can be no early arbitration as to whether the certificate was in accordance with the work done or whether the architect has wrongly withheld payment on the grounds that work executed was not properly executed, or for any other reason.

All the numerous matters which are set out on page 159 of *Building and Civil Engineering Forms* and in *Hudson* 9th Edition, pages 368–9 are not subject to immediate arbitration. The submission on page 159 that 'the words are wide enough to include even simple disputes of valuation on an interim certificate' must therefore be rejected.

Finally, by JCT 80 Article 5.2.3, there can be an early arbitration:

> 'on any dispute or difference under clause 4.1 in regard to a reasonable objection by the Contractor, and clauses 25, 32 and 33'.

JCT 80 clause 4.1 allows the contractor to object to a variation under clause 13.1.2, which alters his working space, working hours or order in which he does specified work.

JCT 80 clause 25 is, of course, the one that deals with extensions of time for completion.

It is significant that there can be no early arbitration under clause 26 'Loss and/or Expense for Delay'. Such claims, if rejected by the architect, must wait until after completion. Clause 32 deals with antiquities, and clause 33 with war.

10.15 Powers of the arbitrator: JCT 80 Article 5.3

The arbitrator in a dispute under the Standard Form is expressly empowered 'to open up, review and revise any certificate' etc. and to determine all matters in dispute 'in the same manner as if no such certificate, opinion, decision, requirement or notice had been given'.

Certain matters are expressly excluded from this power.

Unless it is submitted to immediate arbitration, an architect's statement as to which clause of the contract empowers him to issue an instruction is 'deemed *for all purposes* to have been empowered by the provision of the Conditions specified by the Architect': JCT 80 clause 4.2.

The provision contained in JCT 80 clause 30.9.2 as to the effect of the final certificate overrides it.

So, too, do the provisions contained in the fluctuation provisions, JCT 80 clauses 38.4.3, 39.5.3 and 40.5. The effect of this is that, if there is agreement between the contractor and the quantity surveyor as to the amount of the fluctuations, this shall be binding. This appears so self-evident that it need hardly be stated.

After much controversy, it has now been held that the specific powers that are expressly conferred on arbitrators by the terms of Article 53 are not within the power of any judge and certainly not within the powers of an Official Referee: *Northern Regional Health Authority* v. *Derek Crouch* (1984).

Comments in standard textbooks are therefore now incorrect. No judge has power to open up or review an architect's certificate. This power is specifically given by the JCT contracts only to arbitrators. Judges do not even have this power when the contract (e.g. The Minor Works Agreement) does not give this power to the arbitrator.

10.16 The *lex fori* of arbitrations: JCT 80 Article 5.5

There are in any arbitration three laws which may have application. There is 'the proper law of the contract', that is to say the one by which the parties expressly or impliedly intended to be governed. There may also be the law of the place where the contract is to be performed. Finally, there is the *lex fori*, the law of the place where the arbitration actually takes place. A contract may be in an English form, for the performance of work in Scotland and the arbitration arising out of the contract might be held in Jersey (very sensibly, since the Jersey courts have no power to interfere with arbitrations and any awards made there are registrable and enforceable in other countries. Guernsey unfortunately has introduced all the most ridiculous features of English law.)

Difficulties arose when a construction contract in the JCT 63 English form was performed in Scotland and an arbitration was held in Scotland. At that time there was the possibility in England of an appeal to the High Court by way of case stated but no such procedure in Scotland. The five law lords in the House of Lords were unable to agree as to whether the proper law of the contract was English or Scottish. But they were agreed that the law of procedure of the place where the arbitration took place, the *lex fori*, governed the arbitration and that there was therefore no appeal by way of case stated.

As the result of that case, JCT 80 now provides under Article 5.5:

> 'Whatever the nationality, residence or domicile of the Employer, the Contractor, any sub-contractor or supplier or the Arbitrator, and wherever the Works or any part thereof are situated, the law of England shall be the proper law of this Contract. . . .'

It follows that if work is done in the Channel Islands, the law applicable, the parties agree, is to be that of England. The clause then goes on:

> '. . . and in particular (but not so as to derogate from the generality of the foregoing) the provisions of the Arbitration Acts 1950 (notwithstanding anything in section 34 thereof) to 1979 shall apply to any arbitration under this Contract wherever the same, or any part of it, shall be conducted.'

This is not the place to discuss this in detail, but this part of the clause appears to be self-defeating since the Arbitration Act 1975 makes special provisions for arbitrations which are not 'domestic' arbitrations because one of the parties is resident or domiciled overseas or the subject matter is overseas. Moreover, the parties cannot by contract override the *lex fori* where the arbitration takes place, any more than they can dictate to the English High Court that it shall follow the procedure of Ruritania. Still less can they provide that those Acts of Parliament which expressly do not govern arbitrations held in Scotland or Northern Ireland shall be extended to do so.

Chapter 11

Practical completion, defects liability period, the final account

11.01 Practical completion — what is it?: JCT 80 clause 17.1

There has been a change in the wording of the clause in JCT 80 concerning practical completion. Previously under JCT 63, when in 'the opinion of the Architect the works are practically completed' he was required forthwith to issue a certificate to that effect. It was apparently thought that that expression was ambiguous, in that it did not cover the situation where the works were substantially finished but there were defects. From *Jarvis* v. *Westminster Corporation* (1969), it would appear that an architect was required to issue this certificate even if he knew of existing defects, whereas according to the other House of Lords' case of *Kaye* v. *Hosier & Dickinson Ltd* (1972) the architect was entitled to withhold his certificate until all known defects, except trifling ones, were corrected. The provision now reads in JCT 80 clause 17.1:

> 'When in the opinion of the Architect Practical Completion of the Works is achieved, he shall forthwith issue a certificate to that effect and Practical Completion of the Works shall be deemed for all purposes of this Contract to have taken place on the day named in such certificate.'

This appears to do little to remove the ambiguity.

The difficulty is that the architect's power to order the remedying of defects during the defects liability period is limited to those defects 'which shall appear' during this period. It does not relate to defects which have already appeared and are known to him at the date when otherwise there would be practical completion.

The consequences of the issue of this certificate are so serious that a prudent architect will take steps to ensure that no certificate of practical completion is issued if there are any defects other than trifling ones.

11.02 The consequences of the certificate of practical completion

When the architect issues this certificate:

(1) the contractor ceases to be liable for liquidated damages.
(2) Where JCT 80 clause 22A is adopted the works cease to be at the risk of the contractor and his obligations to insure terminates. Henceforth, they are at the employer's sole risk.
(3) The employer's right to deduct the full retention percentage ends: JCT 80 clause 30.4.1.3.
(4) The defects liability period begins to run: JCT 80 clause 17.
(5) The period of final measurement and valuation begins: JCT 80 clause 30.6.1.2.

JCT 63 clause 30(4)(b) provided that the architect, on issuing the certificate of practical completion, should issue a certificate for half of the total retention monies and that the contractor should be entitled to payment of that within fourteen days from the issue of that certificate. JCT 80 no longer contains such provision. There have in fact been substantial alterations to the provisions regarding retention monies.

A footnote to JCT 80 clause 30.4.1.3 says: 'By operation of clause 30.4.1.2 and 30.4.1.3, the contractor will have released to him by the Employer *upon payment of the next Interim Certificate* after Practical Completion of the whole or part of the works *approximately* one half of the Retention in the appropriate part' (author's italics).

The reader can judge for himself whether that is in fact what the contract says:

30.4.1 'The Retention which the Employer may deduct and retain . . . shall be such percentage of the total amount . . . as arises from the operation of the following rules:

.1.1 the percentage . . . deductible under 30.4.1.3 shall be one half of the Retention Percentage. . . .

.1.2 the Retention Percentage may be deducted from so much of the said total amount as relates to:

work which has not reached Practical Completion . . .

and

amounts in respect of the value of materials and goods included under clauses 30.2.1.2, 30.2.1.3 (so far as that relates to materials and goods as referred to in clause 21.4.1 of Sub-Contract NSC/4 or NSC/4a as applicable);

.1.3 half the Retention Percentage may be deducted from so much of the said total amount as relates to work which has reached Practical Completion (as referred to in clauses 17.1, 18.1.2 or 35.16) but in respect of which a Certificate of Completion of Making Good Defects under clause 17.4 or a certificate under clause 18.1.3 or an Interim Certificate under clause 35.17, has not been issued.'

All this means that, as before, on practical completion half the retention money retained is released to the contractor, but no longer on a special certificate payable within fourteen days but as part of the next interim certificate.

11.03 Provisions regarding partial completion: JCT 80 clause 18

Where with the consent of the contractor the employer takes possession of any part of the works before the certificate of practical completion of the whole has been issued, JCT 80 clause 18 provides that the architect shall issue a certificate giving the approximate value of the works taken over.

So far as that part is concerned, practical completion, with all its consequences [10.02] is deemed to have occurred on the date when the employer took possession — not the date of the certificate of value. Where JCT 80 clause 22A is operated, the risk in respect of that part passes to the employer.

There is to be a pro rata reduction in liability for liquidated damages. This is the ratio not between the total final account figure, which of course will not be known at this stage, but between the contract sum and the approximated certified value: JCT 80 clause 18.1.5.

There is also said to be provision for release of one half of retention monies pro rata in JCT 80 clause 30.4.1.

11.04 The defects liability period: JCT 80 clause 17.2

The defects liability period is named in the Appendix to the Standard Form.

During this period the contractor is required to make good at his own cost any defects which 'appear' in this period, but in this period alone, and 'which are due to materials or workmanship not in accordance with the contract'. But he is not required to do this until the architect delivers to him 'not less than 14 days after the expiration of the period' a schedule of defects.

The effect of these provisions is substantially the same as in JCT 63 clause 15(2) which was discussed by Judge Newey in *Neville (Sunblest) Ltd* v. *Press* (1981).

All defects are breaches of contract, the judge pointed out, and therefore the defects clause amounted to a privilege and right for the contractor who otherwise might have no right to re-enter the site, if he had left; moreover, the employer would otherwise be entitled to employ other people to correct the defects.

The employer was not limited by the defects clause, in that he could sue the contractor in respect of the defects for damages for breach of contract if, for example, he suffered loss by reason of being unable to let or use the premises during this period or subsequently.

In addition to the express provision for the serving of a Schedule at the end of the period, the contract provides in JCT 80 clause 17.3 that:

> '. . . the Architect may, whenever he considers it necessary so to do, issue instructions requiring any defect . . . which shall appear in the Defects Liability Period . . . due to materials or workmanship not in accordance with this Contract . . . to be made good.'

When the contractor has made good all the defects which have been notified to him, the architect has to issue a certificate of completion of making good defects.

In JCT 80 clause 17.2 and 17.3 the words 'at his own cost' are qualified by a sentence in parenthesis 'unless the Architect shall otherwise instruct, in which case the Contract Sum shall be adjusted accordingly'. The meaning to be attached to these words is obscure. The only reasonable construction that can be placed on them is that the architect may instruct the contractor to remedy defects which are his breaches of contract by paying the contractor at the employer's expense. Any architect who issued such instructions is likely to find that the remedial work is done at his expense. The insertion of the word 'and' before the parenthesis makes it clear that the architect has no power, having specified defects, to order the contractor not to remedy them with an adjustment to be made in the final account by way of abatement of the sum due to the contractor. The whole drafting, which repeats JCT 63, is most unfortunate — all the more so because of the provision in JCT 80 clause 30.2.2.1 that any payments made or costs incurred by the contractor under clauses 17.2 and 17.3 shall be included in an interim certificate and not subject to retention.

There is a similar provision regarding the final certificate in JCT 80 clause 30.6.2.11. This seems to be confirmed by NSC/4 clause 14.4 which provides that, if the architect instructs that making good of defects shall not be entirely at the contractor's own cost, 'the Contractor shall grant a corresponding benefit to the sub-contractor'.

This looks as if what is indeed contemplated is that the employer should pay the cost of remedying the contractor's breaches of contract. Where such unbelievably generous employers are to be found is not indicated, and all that can be said is that if an architect were to issue such an instruction without his client's express approval he would most certainly be guilty of negligence. The law does not require an innocent party to pay for the other's breaches of contract.

There also appear to be provisions in NSC/4 whereby a nomination sub-contractor can get paid for remedying defects, and the contractor collect one thirty-ninth discount and a profit mark-up: see NSC/4 clause 14.4, 21.4.2.1, 21.4.2.4.

11.05 Release of balance of retention monies

The contractor appears to be entitled to release of one half of retention monies in the next interim certificate after the certificate of practical completion [11.02].

The balance is released in the next interim certificate after the issue of the certificate of completion of making good defects. There are now no separate certificates for release of retention.

JCT 80 clause 30.1.1.2 provides that the employer is entitled to exercise any right under the contract of deduction from monies due to the contractor in any interim certificate whether or not that interim certificate includes retention monies.

This is subject to qualifications about deduction from interim certificates of monies paid direct to sub-contractors — see JCT 80 clauses 35.13.5.3 and 35.13.4.4.

11.06 Penultimate certificate: JCT 80 clause 30.7

'As soon as practicable' the architect is required to issue an interim certificate which shall include the amounts of sub-contract sums for all nominated sub-contracts finally adjusted.

This is to be issued not less than twenty-eight days before the issue of the final certificate and even though less than a month has passed since the last issue of an interim certificate. There was no such provision in JCT 63.

11.07 Final account: JCT 80 clause 30.6.1

Since JCT 80 is a lump sum contract, the final account is prepared by taking the contract sum and making deductions from that and then making additions to it. For this purpose the valuations and assessments made for interim certificates are ignored. That balance can be in the employer's or contractor's favour. JCT 80 clause 30.8.2 provides that the contract sum, adjusted as necessary in accordance with JCT 80 30.6.2, shall be stated in the final certificate, with the sum of the amounts already stated as due in interim certificates and paid, and the 'difference (if any) between the two sums shall . . . be expressed in the said Certificate as a balance due to the Contractor from the Employer or to the Employer from the Contractor as the case may be'. This sum becomes a debt payable fourteen days from the date of the certificate, subject to the deduction of sums authorised by the contract.

The contractor is under an obligation 'either before or within a reasonable time after Practical Completion' to send to the architect or, if so instructed by him, to the Quantity Surveyor all documents necessary for the adjustment of the contract sums: JCT 80 clause 30.6.1.1.

Table of cases

Note

The following abbreviations of Reports are used:

AC – Law Reports Appeal Cases Series
All ER – All England Law Reports
BLR – Building Law Reports
Ch – Law Reports Chancery Series
KB – Law Reports King's Bench Series
Lloyd's Rep. – Lloyd's List Law Reports
QB – Law Reports Queen's Bench Series
WLR – Weekly Law Reports

Other Reports cited are listed without abbreviation.

Table of statutes

Page references in *italic* indicate that a section is quoted in whole or in part.

Table of JCT contract clauses

Page references in *italic* indicate that the clause is quoted in whole or in part.

JCT 63

Subject Index